Fundamentals of welding metallurgy

Fundamentals of welding metallurgy

H Granjon
Honorary Director of the Ecole Supérieure du Soudage et de ses Applications, Institut de Soudure, France

EDISON WELDING INSTITUTE
1100 Kinnear Road
Columbus, Ohio 43212   614/486-9400

ABINGTON PUBLISHING

Woodhead Publishing Ltd in association with The Welding Institute
Cambridge England

Published by Abington Publishing,
Woodhead Publishing Ltd, Abington Hall, Abington,
Cambridge CB1 6AH, England

English edition, first published 1991, Abington Publishing

French edition, first published 1989, Publications de la Soudure
Autogène

British Library Cataloguing in Publication Data
A catalogue record for this book is available from the British
Library

ISBN 1 85573 019 7

Designed by Andrew Jones (text) and Chris Feely (jacket)
Typeset by BookEns Ltd, Baldock, Herts
Artwork by Institut de Soudure, Paris, France
Printed by Billing & Sons Ltd, Worcester

Contents

Foreword

There is no one better qualified to write a book on welding metallurgy than Henry Granjon. During the course of more than 40 years at the Institut de Soudure, he devoted 30 years of his career to teaching metallurgy at the Ecole Supérieure du Soudage et de ses Applications (ESSA) which trains the engineers and technicians needed by a wide range of industries which use welding, such as mechanical construction, the car industry, naval construction, aerospace, the nuclear industry, etc. There are no industrial fields which do not use welding in the assembly of metal materials.

The teaching of metallurgy at the ESSA was at one time provided in a masterful manner, in all meanings of the word, by Professor Albert Portevin who without doubt inspired the career of Henry Granjon. In 1933, Portevin brought welding technology to maturity by writing a report titled 'The scientific bases of autogenous welding'.

If the scientific basis of this type of assembly was clearly expressed by Portevin, the development, over 60 years, of new welding techniques applicable to materials other than steel demanded that basic welding knowledge be updated. This knowledge was not only thermal, thermomechanical and chemical but was above all metallurgical. Indeed it is possible to weld materials as diverse as stainless steels, zirconium, titanium, uranium, etc, as ordinary iron and steels were welded with oxyacetylene blowpipe 60 years ago. For these materials, it is necessary to know not only their chemical reaction vis-a-vis the atmosphere surrounding the welding, the possible volatilisation of certain additions, but also their metallurgical performance, i.e. the structural transformations caused by the thermomechanical treatments which the welding gives rise to.

So it was the prime interest of Granjon's book to review critically all the welding processes from the most basic to the most modern such as electron beam welding or high energy laser welding. The first chapter is particularly original in proposing a classification of the different welding techniques based on the nature of the energy transfer process to achieve the 'metallurgical continuity' between the two parts to be welded. Five types of energy transfer by gas, electric arc, radiation, mechanical effect and passage of electric current make it possible to distinguish the characteristics specific to the majority of welding processes used, to which must be added less widely used processes such as, for example, those involved in brazing or welding by diffusion in the solid state. Tables 1.1 and 1.2 are particularly instructive and show the wide variety of welding processes available.

The first chapter opens the logical approach to the nine chapters in the book. Following Portevin's study in 1933, the author emphasises the fact that welding presents three successive operations of preparation, casting or solidification and heat treatment.

In a preface limited to a few lines, it is scarcely possible to analyse the whole book because it is so rich in information. It is more useful to draw the reader's attention to the strengths of this book which reflect the author's original ideas, acquired during his long career at the Institut de Soudure and ESSA.

Chapter 2 is an introduction to the thermal and thermomechanical study of welding. It is necessary to understand the distribution of temperature at different distances from the weld bead during and after fusion. Various parameters are involved such as the calorific energy used, the thermal conductivity of the metal to be welded, its volumetric calorific capacity and the maximum temperature reached in the molten pool. Mathematical formulae permit the calculation of the speed of cooling for a maximum temperature θ_m reached at a given point of the metal. The author briefly discusses these mathematical formulae depending on whether products are thick or thin. But very quickly the author shows his interest in describing the different methods used to determine the evolution of temperature in relation to time and distance from the weld bead and also during one or more successive welding passes. This chapter constitutes an introduction which is vital for understanding the structural changes accompanying welding or subsequent cooling which are the subjects of successive chapters in the book. This study of the thermal effects connected to welding cannot ignore their consequences on the deformation and stresses induced by welding.

Chapter 3 concerns the metallographic examination of welds, which is at the base of either quality control of welds, or diagnosis during appraisal of welded parts. Modifications of structure, either in the weld bead, or in the heat affected zone (HAZ) make it possible to determine the method of welding used, for example, the number of passes. The author, despite long experience acquired in the examination of microstructure, has chosen to represent diagrammatically the variations of structure, grain size and constituents arising from the transformation of phases, particularly in the case of steels. In conjunction with the preceding chapter, it is possible to determine in the HAZ the zones corresponding to the temperatures at the beginning and end of austenitisation, A_3 and A_1, on transverse sections at the weld bead. These destructive inspection methods can be preceded by a non-destructive inspection by using the replica method perfected many years ago by P Jacquet, a method which has sadly fallen into disuse but which the Institut de Soudure still uses successfully and profitably.

The following four chapters, 4 to 7, constitute the heart of the book. Development of the fusion zone, its solidification and transformation into the solid phase during welding in periods of heating and cooling constitute the most important stages of the welding operation. It is in these four chapters that the author most demonstrates the depth of his understanding of the importance of metallurgical factors, particularly for steels.

In Chapter 4, the author recalls the importance of the role played by the phenomenon of dynamic recrystallisation under the simultaneous effects of temperature and stress created by welding. This recrystallisation can lead to 'overheat', an exaggerated grain increase also called secondary recrystallisation, which affects mechanical properties and notch toughness in a pure metal not undergoing structural transformation during welding and cooling.

It is in this chapter that the author shows that he has not only acquired long experience in welding of carbon steels. He recalls the problems posed in welding Cr-Ni austenitic stainless steels, aluminium alloys structurally hardened by the addition of copper (duraluminium) or zinc (Zicral) or more stable Al-Mg alloys, at least those in which the magnesium content does not exceed 3%. If the duraluminium type Al-Cu alloys are unusable in welded construction, the same is not true for the Al-Mg and Al-Zn alloys where nevertheless problems of volatilisation of additional elements may arise.

In the same chapter the author examines the phenomenon of the ageing of mild steel associated with the precipitation of Fe_4N in ferrite, of tempering and over-tempering and allotropic transformation of steels which lead to the formation of austenite above temperature A_3 during heating and during its transformation into proeutectoid ferrite and pearlite for the slowest cooling speeds.

This transformation of structure is the subject of Chapter 7 where the author rightly stresses the greater interest presented by the continuous cooling transformation (CCT) diagrams than the TTT diagrams. These determine the kinetics of transformation of austenite into various constituents in the isothermal condition, which is not the case in welding. So CCT diagrams were specially devised for welding. Figure 7.4 is a good example which deserves more comment. Indeed, depending on the cooling speed and composition of carbon steel, these CCT diagrams define the conditions of pearlitic, bainitic or martensitic transformation.

These transformations present characteristics in welding depending on whether the welding is done in one or several passes (on the face or reverse side). Here again, the author demonstrates his teaching qualities by explicit diagrams illustrating, for example, the epitaxial growth of grains in the weld bead from the grains of the parent metal.

Chapter 8 is devoted to the important problem of cold cracking of welds in steel, whose composition of carbon and certain additions like manganese plays the principal role. In fact, the formation of cold cracks in steels results from the combination of several parameters. The first is the hardening which leads to the formation of a large quantity of martensite. This formation of martensite can be estimated by the notion of equivalent carbon calculated by adding to the carbon content the additions content, each assigned a coefficient taking account of the role played by each addition in the hardening of the heat affected zone. Thus, the formula proposed by the International Institute of Welding takes account in this calculation of the Mn, Cr, Mo, V, Cu and Nb content. In practice, it is possible only to retain the value $C + Mn/6$, which must not exceed 0.40. Beyond that, the user must take particular precautions.

The second parameter which favours cold cracking is the hydrogen introduced during welding by the breakdown of water. The third parameter concerns the residual stresses created during cooling of the weld, including those caused by the differences in expansion coefficient of the various constituents of the steel and the expansion accompanying the formation of martensite between M_s and M_f. These facts make it possible to distinguish the localisation of cracks, either transverse or longitudinal at the weld bead, in the HAZ, in the fusion boundary along the length of the weld bead, or in the bead itself. Hence the point of cold crack tests like those in the British CTS test, the Japanese TEKKEN test and the French implant test proposed by the Institut de Soudure. The author makes a pertinent comparison of these three tests to evaluate the susceptibility of a steel to cold cracking not only in relation to its composition but also in accordance with the conditions adopted for welding.

Finally, this chapter examines the evolution of composition of steels for welded constructions which require increasing mechanical performance in terms of strength, ductility and resilience/notch toughness. If the tendency to reduce the carbon content, considered to be the 'welder's enemy' remains, the increase in mechanical properties has been obtained in recent years by additions of low content Nb, giving rise to micro-alloy or HSLA (high strength low alloy) steels in which the increased elastic strength is due both to the fine dispersion of niobium carbon particles and to a very small final grain size due to 'controlled rolling'. 'Dual phase' steels are another solution. The weldability of these steels has much improved because of the low carbon content ($C < 0.10\%$) associated with the formation of bainitic bands instead of lamellar pearlite as a result of quenching between temperatures A_1 and A_3.

Chapter 9 concerns post-weld heat treatments such as stress relieving or tempering, normalising treatments and finally hardening and tempering. The author compares the advantages or disadvantages of stress relief treatments depending on whether they are total or local and depending on the size of the welded assembly. These treatments, carried out in all cases at a temperature lower than that of the A_1 transformation point of steels, do of course have a tempering effect and therefore a not always beneficial influence on mechanical properties by causing the cracking phenomenon called 'reheat cracking'. Hence the reason for specifying the stress relieving mechanism on the basis of temperature and duration of treatment by using the H

'Hollomon' parameter which is introduced into a standard giving the values of H in terms of temperature and time.

Depending on the composition of steels, all these post-welding operations can give rise to reheat cracking as opposed to crack formation during cooling of a structure. This cracking, which HSLA or C-Mn dispersoid steels do not suffer from, appears in certain alloy steels containing carbide forming additions, such as Cr-Mo-V steels. This hot cracking begins at the interface of the weld bead and parent metal and spreads through the HAZ along the joints of the grains of the initial austenite which, during post-weld cooling, has given rise to martensite or bainite. This is a good example of the importance of structural transformations in steels due to the formation of the weld bead and its subsequent cooling. It is another example of the principle which has guided the author throughout his book by demonstrating the interconnection between successive chapters. The transformations of phase which are peculiar to steels occur at all stages of welding operations. Does the same hold true for other materials such as zirconium, titanium and their alloys which also undergo important transformations of structure connected to the allotropic transformations of zirconium and titanium? That will doubtless be the subject of another book inspired by this one.

Finally the last chapter constitutes a critical account of destructive and non-destructive welding tests. Granjon rightly points out that non-destructive tests developed for checking welds have been the basis for much progress made in non-destructive tests applied to non-welded materials.

In addition to destructive traction, bending, tearing tests (for spot welds) and tests based on the (supposedly linear!) fracture mechanics, the author puts more emphasis on non-destructive tests – liquid penetrant, magnetic particle testing, radiography, ultrasound and acoustic emission. The last figure in this book reflects again the author's preoccupation with metallurgical factors. This figure (10.14) on the one hand shows the relative efficiency of the five non-destructive tests for detecting cracks (emergent or not), cavities, inclusions and lack of fusion (emergent or not). But this detection efficiency is influenced to a greater or lesser degree by the metallurgical factors of grain size and orientation, variation of magnetic permeability and X or γ absorption coefficient of the different phases (as with austenitic ferritic stainless or austeno-ferritic steels) and finally the nature of heat treatments.

This final figure illustrates, if such illustration were necessary, the author's constant preoccupation to demonstrate the importance of metallurgical factors on the quality of welds in order that they shall best fulfil their function:to ensure the joining of two metal elements without provoking too great an effect on the mechanical properties of the assembly compared with those of the parent metal. To distinguish clearly the role of the different thermal, mechanical and metallurgical parameters during fusion, solidification of the weld bead and heat treatments concurrent with or subsequent to welding, the author had to have behind him a long career as engineer, researcher and teacher. The risk in fulfilling this objective was that of compiling an encyclopaedic book and producing an exhaustive compilation of results from which only the specialist could derive precise information.

Finally, the teaching qualities of Granjon are illustrated by the large number of figures which are more than diagrams and which are more informative than micrographs of microstructures. For example, each figure representing sections of multipass weld beads implies that the author has himself often put his eye to the microscope so that the image he proposes shall conform to reality.

In conclusion, this book should appeal to a wide audience, not necessarily expert in welding, wishing to know more about the different welding processes applicable to the most diverse materials provided they understand the importance of metallurgical factors. So welding has ceased to be an 'art' but has become an assembly process which has a solid scientific basis. Granjon's book is there to convince us of this thanks to his enthusiasm and broad metallurgical culture.

P Lacombe
Member of the Academy of Science

Preface

The title of this book 'Fundamentals of welding metallurgy', was inspired by the title of a text 'The scientific bases of autogenous welding' in which Professor Albert Portevin, as early as the 1930s, set down the foundations of what was later called the metallurgy of welding. In choosing this title, the author first wanted to pay homage to the memory of his old teacher, who created the professorship of metallurgy at ESSA. He also wanted to summarise the motivation behind the writing of these pages, namely the desire to know and understand the essential facts of the metallurgical phenomena which the various phases of welding use or generate. The originality of the author's approach compared with traditional so-called 'general' metallurgical teaching lies in the particular characteristics assumed by these phenomena.

Such an objective can be sought at different levels, and developed in more or less depth; the corresponding instruction must be adapted to the knowledge and functions of those at whom it is directed, so that each person can, at his level of activity, derive practical benefit. This is the spirit in which the author did his best not to go too far beyond the limits of popular knowledge. In these conditions, part of his readership will nevertheless have to refer to some basic metallurgical books for some knowledge; conversely, other readers, appetites whetted, will be tempted to refer to books or articles explaining or developing to a higher level the deliberately limited information contained in the chapters of this book.

It was in thinking about the use which could be made of this book by teachers that the author tried to make the chapters as independent as possible from each other, so that the overall plan could, in part or as a whole, be a course or a training session, short or long. Each chapter can be reduced or expanded in accordance with the needs of the audience.

Some references which were deemed necessary have been included, but in as general a manner as possible. To illustrate phenomena, the performance of materials most frequently met in common practice is given by way of example. Therefore there are no chapters specially devoted to particular materials, except with regard to steels and the consequences of hardening, which call for specific precautions. Such could have been the case, given a different framework, but at the price of inevitable repetition.

The figures comprise exclusively diagrams or representations, most of which have been developed and used by the author for his own teaching and some of which, to his great satisfaction, have already passed into the literature, under the pen of old pupils or readers of previous articles. Photographs, of debatable interest if documents are presented in small numbers and with incomplete annotations, or which overload the text if they are numerous and suitably annotated, were deliberately excluded. All the same, lively teaching of welding metallurgy could not be undertaken without the support of metallographic articles and, better still, direct examination of typical samples.

The bibliographic references have been reduced to a minimum by limiting them to documents directly mentioned in the text. Indeed, all the subjects dealt with are covered by such a huge bibliography that an exhaustive and annotated list of each one would in itself constitute a considerable book.

Finally, the report includes numerous comments or remarks which have been placed at the end of each chapter so as not to distract the attention of the reader who can, to begin with, make do with a cursory reading returning later for more attentive study.

In concluding this preface, the author sincerely thanks Professor Paul Lacombe, Member of the Academy of Science, who was kind enough to write the foreword to this book, a gesture which bears witness to the interest he has always shown in contributing to fundamental research for the development of technology and welding in particular. He also thanks his old colleagues and associates from the Institut de Soudure, without mentioning them by name for they are many; they will each easily recognise the part, direct or indirect, they have taken in the production of this book. Finally, the author acknowledges all his old French and foreign pupils and students, whose sustained attention and, in some cases, insistence, persuaded him to write this book, which he dedicates to them, in remembrance of the hours spent together.

H Granjon

1

Metallurgical presentation of the general welding processes and characteristics

Definition of welding: classification of processes

Metallic continuity

Among the numerous and varied definitions of welding supplied by technical literature for the last 75 years, the most obvious for inclusion in a book on welding metallurgy is the one which calls upon the idea of continuity, as found in ISO standard R 857 (1958) and also in the International Electrotechnical Vocabulary (term no. 40-15-005). According to that definition, welding is an operation by which continuity is obtained between parts for assembly, by various means which shall be examined later. The motto which appears on the coat of arms of The Welding Institute says only 'e duobus unum', that is 'from two they become one'.

This definition, valid for any material, including plastics, applies fully to metals and alloys, insofar as the operation consists of establishing metallic continuity between the parts to be assembled. At the macroscopic level, metallic continuity implies the absence of any non-metallic material between the joined elements once assembly is complete. In this, bonding by welding is different from that achieved by rivetting, clamping or even gluing. Continuity thus defined does not imply homogeneity of chemical composition through the joint. A weld can be homogeneous (for example, a mild steel weld with filler metal of the same type) or deliberately dissimilar (for example, a cast iron weld with aluminium bronze filler metal). In the bond area unintentional heterogeneity resulting from the welding operation itself (for example decarburisation with steel welding) can also be seen.

Closer examination of what happens with regard to crystal structure reveals more on the nature of metallic continuity achieved by welding.

Prior to the operation, the atoms of each part to be assembled are gathered into two distinct groups. After welding, metallic continuity shows itself in that these two groups have joined together to form a single entity, without the interposition of any extraneous atoms (if, in the interests of simplicity, we consider the case of a weld without filler metal). This is what is illustrated in Fig. 1.1, where in assembly AB resulting from the welding operation, the broken line delineates what is called the fusion boundary, on either side of which are the atoms originating in the initially separate groups A and B. The main characteristic of the welding operation lies in the fact that this fusion boundary contains no metallurgical discontinuity in the atoms present. Every welding process must therefore include a process of elimination of non-metallic extraneous atoms which could prejudice the bond.

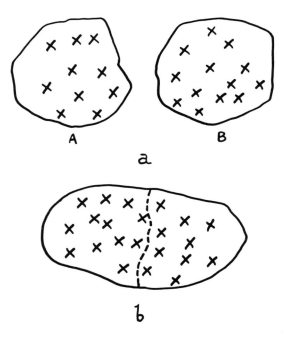

1.1 Nature of metal continuity: constituent atoms of parts A and B, belonging initially to two separate assemblies (a), forming a single assembly after welding (b).

Mechanisms for obtaining metallic continuity

In actual fact, metallic continuity is not defined solely by whether the atoms present are metallic or not. Indeed, contrary to what is suggested by Fig. 1.1 (deliberately simplified), account should be taken of the fact that the constituent atoms of metal parts A and B which are to be joined together are arranged in a well defined geometric pattern, in which the orientation varies from one grain to the other. So, given the orientation of the atoms in the constituent grains of the parts for assembly, there is a case for examining the mechanisms by which continuity is established. To do this, we will first of all consider what happens in a grain or, more precisely, we will do our reasoning as if the parts for assembly were each made up of a single grain, called a monocrystal. We shall distinguish three mechanisms, illustrated by Fig. 1.2, 1.3 and 1.4.

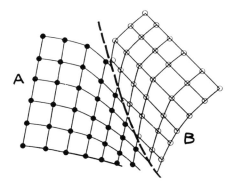

1.2 Achieving metal continuity by bringing together atoms obtained by cold deformation. The crystalline lattice is distorted at the fusion boundary (broken line).

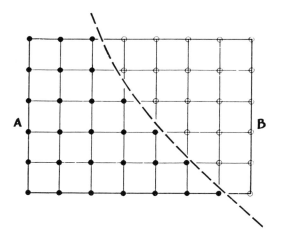

1.3 Achieving metal continuity by combined effect of deformation and recrystallisation which creates a common crystal lattice on either side of the fusion boundary.

A first mechanism for achieving metallic continuity consists, in the solid phase, of bringing together metal atoms sufficiently closely to ensure that, on the one hand, extraneous atoms are excluded and, on the other hand, aided by mutual attraction, that the rapprochement or bringing together is permanent. This mechanism can occur in the hot or cold state, by means of deformation. For example, this is the

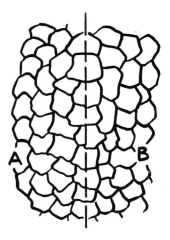

1.4 Achieving metal continuity by hot deformation (recrystallisation). On the microstructural scale grains common to the assembled elements arise from nuclei produced at the interface.

mechanism in ultrasonic welding or cold pressure welding, and also in butt resistance or friction welding. If the bringing together of atoms is caused by cold deformation, the crystalline structure is adversely affected after welding, because the deformation causes work hardening (Fig. 1.2). If deformation occurs in the hot state, metallurgical continuity is completed by recrystallisation, which establishes a common orientation on both sides of the interface (Fig. 1.3). In the microstructure, this process results in a common grain texture in the assembled elements. These grains are produced by recrystallisation from single nuclei, products of both elements (Fig. 1.4).

A second mechanism involved in establishing metallic continuity is diffusion, that is the passage of atoms from part A into part B and vice versa (Fig. 1.5). If the material remains in the solid state on either side of the interface, diffusion takes place in both directions (heating and pressure are vital for this to occur). This is the phenomenon which characterises the type of welding said to be 'by diffusion' but diffusion is not involved alone in this process, for there is also recrystallisation, under the simultaneous effect of heat and pressure. If the interface separates a material which is liquid at the moment of welding and a material which remains solid because it is less easily melted, diffusion permits the liquid atoms to pass through this interface and lodge themselves in the crystal system of the solid metal, by creating a very thin area of alloy over the length of the interface. In the other direction, atoms from the solid material can pass into the liquid, but they become diluted throughout the entire mass, causing only an imperceptible change in its chemical composition. This is how the processes of brazing and braze welding, which are characterised by the solidification of a liquid phase on a solid phase whose melting point is higher, are described. It should also be noted that, even when not acting as a principal factor in the process of achieving continuity, diffusion is involved in

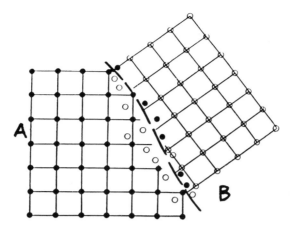

1.5 Achieving metal continuity by solid phase diffusion across the interface (diffusion welding). If B is liquid, the A atoms become diluted (brazing and braze welding). Or the A and B atoms in the solid state may join together to supply a liquid phase AB at the interface.

fusion welding whenever there is a difference in chemical composition on either side of a bonding zone, in other words frequently.

A third mechanism results from the solidification of a liquid B on a solid A from which it originated, or at least participated in the fusion. In this case, metallurgical continuity is provided by epitaxy, in which solidifying crystals, forming at the interface, orientate themselves on the crystals of the solid parent metal. Thus the crystal structures in the grains and in the boundaries between them are common to the parent metal metal which has remained in the solid state and to the newly solidified molten metal. As for the microstructure, this phenomenon (Fig. 1.6) means that the grain boundaries of the parent metal cross the interface, which has become the fusion boundary, without discontinuity. This is seen in all welding processes in which fusion occurs, with total or partial participation of the parent metal in the development of the weld metal zone, i.e. processes such as gas welding, arc welding, resistance welding, spot welding and many others, which we shall later call 'autogenous'.

Thus, with regard to crystal structure, three mechanisms contribute, separately or together, to the establishment of metallurgical continuity, namely the bringing together of atoms by mechanical force, diffusion and creation of a common crystal orientation by recrystallisation or epitaxy.

Circumstances for achieving metallurgical continuity

When we consider all the welding processes in order to understand the circumstances by which metallurgical continuity is achieved by one or more of the above mechanisms, we arrive at an extremely simple classification,

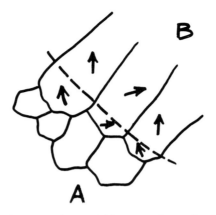

1.6 Achieving metal continuity by epitaxy during solidification at the interface between liquid B/solid A. The identity of grain orientation on either side of the fusion boundary results in the extension of their boundaries across that zone.

because all processes involve one of the three operations below in the welding operation:

a) Liquid/solid interface: the processes in which the bond is obtained by solidification of a liquid phase on contact with a solid parent metal.

b) Solid/solid interface: this class contains the process ensuring metallurgical continuity from solid state contact between the parts for assembly.

c) Vapour/solid interface: this is condensation of a filler in the vapour state on to a parent metal which remains in the solid state. This relatively recent process is applied in some cases of brazing or surface coating.

The first two classes above in turn break down as follows:

a.1 Liquid/solid interface with fusion of parent metal, i.e. with exclusive or partial participation (use of a filler metal) of the parent metal in the development of the fusion metal. These are welding processes once called 'autogenous', in the etymological sense of the word and seldom used nowadays, because for historical reasons it brings to mind gas welding, particularly in the German and Russian languages.

a.2 Liquid/solid interface without fusion of parent metal, i.e. without participation of the parent metal in the development of the weld metal. This heading concerns brazing and braze welding.

b.1 Solid/solid interface, obtained by prior formation of a transient liquid or viscous phase, itself arising from the parent metal, which is eliminated during the operation, either globally (flash or friction welding), or step by step (explosive welding).

b.2 Solid/solid interface, with direct bond in the solid state, by one of the processes described above (cold or hot deformation).

These various interface situations and the type of bond which results from them are achieved through three exclusive or combined actions, namely:

– use of a filler metal (m),
– application of pressure to the parts to be assembled (with consequent overall or local deformation) (p),
– intervention caused by temperature variation at the interface (t).

The left side of Table 1.1 shows the different combinations of these conditions currently in use for each type of interface leading to the creation of metallic continuity.

Classification on the basis of type of activation energy of interface

Having listed the metallurgical elements in the classification of welding processes, these processes still need to be distinguished by the nature of the energy used to obtain the action necessary to establish metallic continuity at the boundary. It was for this purpose that two working groups from the International Institute of Welding,[2] with a view to classifying literature on welding, for documentary purposes, drew up a list of nine processes of energy transfer at the boundary*, permitting a finer distinction than previous classifications. This list, reduced here to 6 transfer processes referenced A to F in the interests of simplicity, is used in the right hand section of Table 1.1. Combined with the indications relative to the interface and the conditions for obtaining it, this list made it possible to draw up the list in Table 1.2, completed below by a few comments.

A: Gas transfer: This group includes the processes in which the energy necessary for bonding is obtained at the interface by a calorific exchange between the parent metal – and filler metal where applicable – and a hot gas or gas mixture. We find here the processes using a flame, whether it is gas fusion welding (a.1.1 A or a.1.3 A), brazing or braze-welding (a.2.1 A), or gas pressure welding (a.2.2 A). We must also classify under this heading plasma welding (a.1.1 A or a.1.3 A), because in this process, energy transfer is provided by the ionised gas jet which constitutes the plasma[3], as was the case in atomic hydrogen welding, a process now outdated.

B: Electric arc transfer: This is the most important mode as regards the number of variations and tonnage of products welded. In most processes, the electric arc causes fusion of the parent metal and the bond occurs by solidification of the molten metal on to the parent metal, either without filler metal (e.g. a.1.1 B, TIG welding, i.e. with a non-fusible electrode) or especially with filler metal, the heading a.1.3 B grouping together all the variations of electric arc welding (with electrodes or electrode wire, with gaseous or solid fluxes, or under slag). The electric arc can also be used to achieve a bond by liquid-solid contact with a filler metal less easily melted

---

*Doc IIS VI-582-86: List of welding and allied processes, with their definitions, classified according to energy source

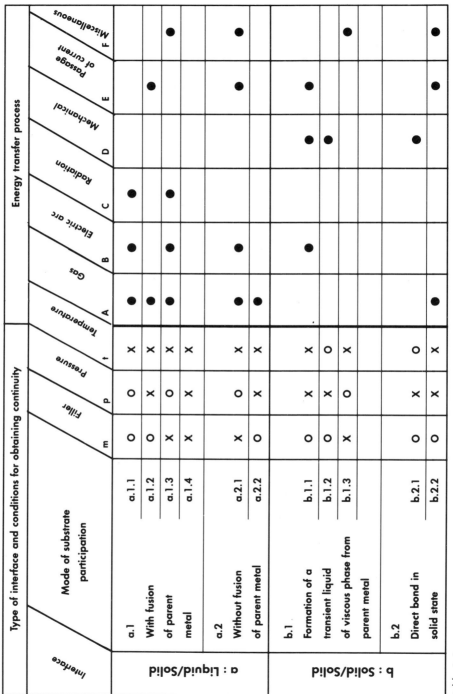

Table 1.1

Type of interface and conditions for obtaining continuity						Energy transfer process						
Interface	Mode of substrate participation	Code	Filler (m)	Pressure (p)	Temperature (t)	Temperature (A)	Gas (B)	Electric arc (C)	Radiation (D)	Mechanical (E)	Passage of current (F)	Miscellaneous
a : Liquid/Solid — a.1 With fusion of parent metal		a.1.1	O	O	X	●	●	●				
		a.1.2	O	X	X	●	●			●		
		a.1.3	X	O	X	●	●	●				●
		a.1.4	X	X	X							
a.2 Without fusion of parent metal		a.2.1	X	O	X	●	●					
		a.2.2	O	X	X	●				●		●
b : Solid/Solid — b.1 Formation of a transient liquid of viscous phase from parent metal		b.1.1	O	X	X		●		●	●		
		b.1.2	O	X	O				●			
		b.1.3	X	O	X							●
b.2 Direct bond in solid state		b.2.1	O	X	O				●			
		b.2.2	O	X	X	●				●		●

8

a.1.1 A	Welding without filler metal (flame or plasma)
a.1.3 A	Welding with filler metal (flame or plasma)
a.2.1 A	Gas brazing and braze welding
a.2.2 A	Gas pressure welding
a.1.1 B	Arc welding without filler metal with non-fusible electrode (TIG)
a.1.3 B	Arc welding with non-fusible electrode (TIG) with filler metal
	Arc welding with electrode or submerged-arc welding and vertical electroslag welding
a.2.1 B	Arc braze welding
a.1.1 B	Rotating arc welding
a.1.1 C	Electron beam welding
or	Laser welding
a.1.3 C	Solar energy or arc image welding
b.1.1 D	Friction welding
b.1.2 D	Explosion welding
b.2.1 D	Cold pressure welding
	Ultrasonic welding
a.1.2 E	Resistance spot welding
	Resistance brazing
b.1.1 E	Flash welding – high frequency induction welding
b.2.2 E	Resistance butt welding
	Mean frequency induction welding
a.2.1 F	Brazing
b.1.3 F	Brazing – diffusion
b.2.2 F	Diffusion welding
a.1.3 F	Aluminothermic processes
a.2.0 F	Hand soldering
b.2.2 F	Forge welding

Table **1.2**

than the metal substrate. This is arc braze welding (a.2.1 B). Finally, with the arc between the parts to be welded, one can, using pressure, eliminate the liquid phase to achieve a bond in the solid state. This is the case with rotating arc welding (a.1.1 B), used for achieving circular bonds.

C: Radiation transfer: In this group we have the use of electron beams or lasers or, even though anecdotal at present, the use of high energy source images (solar energy or electric arc). All these processes use only fusion; so they are found only under heading 1.1. Electron beam welding is most often used without filler metal (a.1.1 C), but it includes some applications requiring the use of such metal (a.1.3 C). These applications are more rare in laser welding.

D: Transfer by mechanical effect:[4] Processes using energy resulting from a mechanical effect (deformation or friction) as sole activation energy all belong in category b

9

because they effect a bond in the solid state, either by the intermediary of a transient liquid or viscous phase, as with friction or explosion welding (b.1.1 D and b.1.2 D respectively) or directly, in the solid phase, as is the case in cold pressure or ultrasonic welding (b.2.1 D).

E: Transfer by passage of an electric current: Under this heading we find the processes in which an electric current is passed through the parts to be assembled, generating heat by the joule effect where these parts are in contact and inside them. This implies the intervention of pressure. It is the case with resistance spot welding (a.1.2 E) where the bond is made in the liquid state, flash welding and high frequency induction welding (both classified in b.1.1 E), where the bond is made in the solid state after expulsion of a transient liquid phase; and finally, it is the case with resistance butt welding (b.2.2 E) where the bond occurs directly in the solid state.

F: Miscellaneous processes: A small number of processes escape classification on the basis of energy transfer process, which is not specified. This is the case with brazing (a.2.1 F) or diffusion-brazing, where filler metal starts by melting then diffuses into the parent metal (b.1.3 F), or diffusion welding (b.2.2 F) where the bond takes place in the solid phase, under pressure and with a non-specified calorific input.

Also, the list of energy transfer processes in columns A to F is not exhaustive and, under the heading 'miscellaneous processes', we can mention, for example, aluminothermic welding (a.1.3 F) and older processes such as hand soldering (a.2.0 F) or forge welding (b.2.2 F).

Allied techniques

Techniques called 'allied' (in relation to welding) are those which, pursuing an objective distinct from that of welding, use energy sources and products similar to those which are used in welding. Independently of building-up, sometimes considered as an allied technique, but which is only one particular application of welding, some of these techniques will be noted in this book, insofar as they provoke or exploit one or other of the phenomena we are studying.

In thermal cutting, sectioning of parts for cutting is obtained by fusion, either directly by means of an arc plasma which supplies the necessary local calorific energy, or indirectly by means of heat from combustion maintained by an oxygen jet. So we talk about 'plasma cutting' in the first case, 'flame cutting' in the second. This latter applies only to ferrous alloys whilst the first also concerns non-ferrous metals. From the point of view of the phenomena we have already mentioned, thermal cutting of steel means, in the case of flame cutting, a chemical evolution at the level of the cut and in all cases a transformation into the solid phase leading to a metallurgical modification and tenso-thermic effect.

Some heat sources used in welding, whether flame, plasma or electric arc, because of their specific power, are used in thermal spraying. The process consists of coating the surface to be treated, whether or not it is metallic, by forcefully spraying on particles of filler metal, obtained by fusion at the heat source. When the filler is metallic (metallising), the process raises chemical problems of protecting the metal during fusion and transfer. The metal applied, strongly work hardened, is the seat of residual stresses, which justifies the precautions which have to be taken to ensure adhesion to the support material.

The localised nature of some heat sources used in welding means that they are also used for treatments such as hardening of steels, whether it is a question of surface hardening (by flame, induction or laser, processes in which rapid cooling is involved) or even more localised though still superficial treatments, using electron beams and especially lasers.

Finally, among allied techniques essentially applicable to steel we can mention those which use a flame, either for forming, or for straightening. In forming, the flame is used to bring the metal to red heat locally; it is then malleable and can assume the shape imposed on it by means of a suitable tool. Flame straightening is achieved by means of local heat which, by impeded expansion causes plastification of the metal and its shrinkage during the rapid cooling which follows gives rise to the desired reverse-deformation.

General characteristics of the welding operation

Comparison with other operations

In a remarkable study which marked an epoch, Portevin* had already, in 1933, identified three aspects in order to categorise the welding operation in comparison to other, at that time better known, metallurgical operations, namely:

– a preparatory operation (at least for processes involving fusion) since, to the extent that we shall discuss in greater detail, the phenomena involved are comparable, apart from scale, to those which in steel for example, produce the desired chemical composition prior to casting.

– a casting operation (still for the same processes) for, after melting into the weld pool, the fused metal solidifies in the same way, at first analysis, as metal which solidifies in the mould into which it has been cast.

– a heat treatment operation, at least for processes involving the use of a heat source which heats the parent metal and causes it to undergo a heat cycle, modifying it in the same way as industrial heat treatment modifies the treated item in the manner desired.

---

* A Portevin: Scientific bases of autogenous welding – Bulletin de la Societé des Ingenieurs Soudeurs, No. 24, 1933, pp 902–25

Indeed, more extensive thought about industrial operations from this reference has led to the definition of welding characteristics compared with these operations, and consequently to the establishment of the limits of this reference. These characteristics themselves boil down to a character common to all welding processes, namely localisation of the operation, which necessitates the use of a powerful heat source to ensure the necessary speed of operation.

From the point of view of development of the molten pool, localisation implies that it takes place progressively and quickly (except in the case of spot welding). This characteristic excludes the state of equilibrium for the chemical reactions which occur and which, moreover, take place at temperatures which, depending on the process, can greatly exceed the fusion temperature. So our conditions are not those of industrial preparation.

As a casting operation, welding is distinct from standard foundry in that the development and therefore solidification of the molten pool occurs step by step, hence the characteristics of the solidification structure of single pass welds, which we will need to examine and compare with the structure of multipass welds. Also, in the more general case of fusion welding, molten metal joins with the parent metal at the moment of solidification, as a result of the intervention of the parent metal, part of which melts and mixes with the filler metal which has melted, thus triggering evolution of the chemical composition of the molten pool. This is the dilution phenomenon and we shall examine its consequences later.

Finally, as a heat treatment, welding is distinguished from the majority of industrial heat treatments by the consequences resulting from the necessary localisation and power of the heat source. Hence in most cases, a short heat cycle, an uneven temperature distribution and a high maximum temperature, reaching and exceeding the fusion temperature.

The three comparisons proposed by Portevin, if they indicated the path to follow, also led to the exploration of fields in which there was no interest before the study of welding metallurgy. But it was not possible to hold to strictly metallurgical studies, for account also had to be taken of the characteristic welding operation temperature gradient, because of the deformation and residual stress it generates.

Finally, to present the general characteristics of the welding operation, three aspects have to be examined:

– the thermal aspect
– the chemical aspect
– the thermomechanical aspect

But we should not lose sight of the fact that the purpose of a weld is to join together the elements of a construction. The continuity thus established gives the welded construction a particular character to which we can also pay some attention.

Thermal aspects

Most welding processes are characterised by the involvement of a heat source, usually mobile, sometimes stationary (for example in spot welding) and the nature and specific power that that source can achieve determine the practical possibilities of utilisation. An interesting comparative representation of these possibilities is given by Rykaline* in the form of what he called the 'heating stain' (Fig. 1.7); the extent of this representation shows, for the power involved in the use of various heat sources, the greater or lesser concentration which characterises these sources. To avoid erroneous interpretation of Rykaline's representation, we must point out that the diagram is given in logarithmic co-ordinates, so that the 'heat stain' relative to the flame, which appears small on the graph, is in reality of the order of a centimetre, whilst the one for electron beams which appears extended on the graph is of the order of a millimetre.

Also, at the time when these data were published, processes such as electron beam welding were still not in use for energy of the order of several dozen or even several hundred kilowatts, as is the case today. So we can no longer attribute a quantitative character to this representation. Be that as it may, Rykaline's representation illustrates the localisation which is characteristic of all welding processes and from which, to various degrees, the following consequences result:

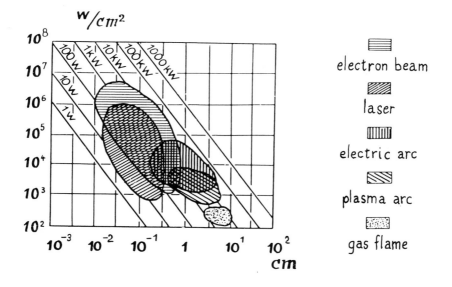

1.7 Extent of the 'heat stain' (Rykaline), of the various welding processes in relation to thermal flux and source rating.

---

*N N RYKALINE: Energy sources used in welding – Soudage et Techniques Connexes – No. 11/12, 1974, pp 471–485.

a) Welding includes rapid heating, in many cases more rapid than that corresponding to the solid phase state of equilibrium of the alloys concerned. If we wish to foresee or interpret the metallurgical effects of welding, it is necessary to take account of the speed of heating, in particular when we wish to simulate the welding heat cycle to reproduce resultant structures.

b) The effect of rapid heating is accentuated by the brevity of time spent at maximum temperature, temperature which is close to fusion temperature. The result is that, from the metallurgical point of view, some compensation occurs between the temperature rise achieved, on the one hand, and the brevity of time spent at that temperature, on the other hand. Given identical chemical composition, this is the explanation of the difference which can be seen between the characteristic structures of welds obtained by different welding processes, depending on the relative brevity of duration at high temperature which characterises them. This compensation effect can also be observed in the liquid phase, for example with regard to volatilisation losses in the fusion zone.

c) Finally, the temperature gradient resulting from localisation of heating gives rise to a generally high cooling speed, itself a function, for each material, not only of the welding process used, but also of the procedure applied when using this process.

In conclusion, for the reasons set out above, any metallurgical analysis of the effect of a welding operation must be preceded and explained by a detailed thermal study of that operation; this study is the subject of Chapter 2.

Chemical aspect

With a few differences due to the brevity of the operation, we have been able to compare welding, at least fusion welding, with a dynamic operation, on the basis of the evolution produced in the chemical composition of the weld metal.[5] We can mention now, before studying them in greater detail in the chapter devoted to development of the weld metal zone, the causes of chemical composition evolution in weld metal, either by the performance of its constituent elements, or by contact with the surrounding medium (gaseous or liquid), or in the presence of the solid medium constituted by the parent metal itself.

a) With regard to the actual weld metal and its constituent elements, the high temperature achieved during the operation can be the origin of element loss by volatilisation, a loss which has to be compensated for by additions or by a suitable procedure. Also, while the metal is in the liquid state, chemical reactions are liable to occur between elements present, whether they come from the parent metal or from the filler metal. Such reactions result in an evolution of mechanical or other properties, and possibly a change for the worse in density if insoluble components are formed or if gases are given off.

b) At the same time, the weld metal is liable to be chemically affected by the surrounding gaseous medium, if such a medium exists. Originally, it was the desire to protect weld metal against the harmful action of oxygen and nitrogen in the air which led to the preparation of coatings and fluxes, which produce a liquid slag which separates the weld metal from the atmosphere, then the use of fully protective gases such as argon, or partially protective such as carbon dioxide gas or mixtures of that gas with argon.

In fact, the gases listed above are not the only ones which have an effect, and we shall in particular have to examine, from other points of view, the role of hydrogen in welding steel or other alloys.

c) The search for protection of the molten pool against the surrounding atmosphere, together with a means of incorporating elements liable to intervene, has led to the development, already mentioned, of fluxes and electrode coatings which, at the moment of fusion, produce a liquid phase called slag which reacts on contact with the molten pool and causes its composition to evolve. So there is indeed some change with time, but as mentioned above, it is a rapid operation, outside the normal conditions of equilibrium.

d) Finally, we have already mentioned the fact that on contact with the solid parent metal which serves as a backing, a more or less significant proportion of the parent metal is incorporated, by its own fusion, into the weld metal. Hence the definition of dilution, whose value depends on the one hand on the welding process, by the penetration which is peculiar to it, and on the other hand, on the procedure for each process.

The four general chemical composition evolution mechanisms during welding which have been mentioned briefly will be examined in more detail in Chapter 4.

Thermomechanical aspect

The localised nature of the welding operation and, where necessary, passage through the fusion state to solidification involves two important physical processes, namely solidification shrinkage and thermal expansion which act simultaneously to generate deformations and stresses at welded joints.

Prevention of the overall deformation caused by welding and associated with shrinkage on solidification and cooling poses technological problems outside the scope of this book. But it should be noted that prevention of deformation results in additional residual stresses. The remedy for deformation therefore, whenever possible, is to look for a procedure which reduces it, rather than a method of restraining the parts for assembly.

As for residual stresses which arise in locally hot (i.e. plastically) deformed zones, these are created during cooling, affecting metal which has become elastic again, which gives them a maximum value equal to the yield stress of the metal. Also, but this concerns only some particularly hardening steels, it is advisable to take account of the stresses associated with martensitic transformation.

We shall return later to the problems posed by deformation and residual stresses and their relief (Chapters 2 and 9). But it should be emphasised now that they are present as a matter of course because of the thermal characteristics of welding. We could not therefore, when establishing a welding programme, content ourselves with a purely metallurgical appreciation of the effects of such a programme. It is absolutely essential to evaluate these effects in the presence of the stress fields which it creates (for example by cracking tests).

Constructive aspect

The question raised here, in the context of general welding characteristics, is that of knowing whether the continuity created by welded joints between elements assembled to build a construction implies any particular requirement with regard to the parent metal, independent of those resulting from the direct metallurgical and thermomechanical effects of welding.

In effect, the continuity provided by (supposedly sound) welds gives the construction a monolithic character which is not without risk. It is easy to understand that, if, at some point of the construction and for some reason not necessarily associated with weld quality, a break is initiated, no discontinuity is likely to halt its development if the parent metal or the welds themselves do not resist it. To illustrate this we need only remember the spectacular breaks which have in the past affected welded constructions such as boats or bridges and the use which was made of lines of rivets to create obstacles (called at the time 'crack arrestors') to prevent such occurrences.

Consequently, at least in the case of steel, the metallurgical study of welding, which made it possible to interpret and master phenomena originating at the welded joint, should be complemented by an in-depth understanding of the behaviour of the parent metal to enable complete mastery of the problem of weldability to be achieved.

General conclusion: what is welding metallurgy?

The above brief account of the general characteristics of the welding operation has shown that metallurgical continuity is obtained at the cost of conditions which are not without effect, metallurgically and thermomechanically, at and beyond weld joints. These effects and their consequences vis-a-vis the behaviour of those joints and of the constructions containing them, during welding and in service, are at the bottom of what is by common agreement called 'weldability'. The aim of welding metallurgy is to study these effects and interpret their consequences, with a view to providing a foundation for the requirements concerning both products described as weldable and the development of procedures relative to the application of welding processes. In everyday language, we could say that the study of welding metallurgy consists of paying attention to what the metal 'thinks' of the welding operation and the memories it retains of it, because weldability depends on those 'memories'.

Such a study of welding metallurgy cannot be tackled in ignorance of metallurgy in the general meaning of the word. That is why this book includes information which should enable the reader to grasp the characteristics of a phenomenon when it is involved in welding.

Chapter 1 **Comments**

1 At this point, we should specify what we call cold and hot deformation, in metallurgy, and hot and cold pressure welding. In metallurgy, we say that cold deformation occurs at a temperature low enough to cause work-hardening, i.e. a displacement of atoms which remains after the force which provoked it has ceased. Deformation is called hot when it occurs above a certain temperature, called recrystallisation, from which the crystalline structure regenerates spontaneously. In pressure welding, we talk of 'cold welding' when no heat source is directly involved to facilitate recrystallisation and 'hot welding' when a heat source is directly involved. The incidence of recrystallisation in welding is dealt with and developed in Chapter 6.
2 'Scientific and Technical Information' Study Group and Commission VI 'Terminology'.
3 Such is not the position of the IIW document which places plasma under heading B (Electric arc).
4 The IIW document specifies that energy is obtained by 'displacement of a mass'.
5 This does not exclude the possibility, for certain welding processes and certain materials, of superficial chemical reactions in the solid phase, for example, in the case of reactive metals or alloys such as titanium.

2 Thermal and thermomechanical study of welding

The thermal study of welding

The wide variety of methods of producing the calorific energy used in the various welding processes means there are big differences between those processes as regards their thermal characteristics and the consequences of those characteristics. Leaving aside the considerations relative to the distribution of the energy at the heat sources,[1] for they belong to the field of physics, we will limit our thermal study of welding to what concerns the molten pool – if there is one – and the parent metal which remains in the solid state.

The purpose of a thermal understanding of the welding operation is first of all to permit as comprehensive an interpretation as possible of the metallurgical phenomena generated by that operation and then to plan those phenomena to take account of them in choosing welding processes and procedures. This planning relies on data permitting effective simulation of the planned operation (i.e. reproduction on test pieces of the effects of that operation), or even, on an analytical simulation[2] by means of suitable programmes, themselves based either on a sufficient number of experimental results, or on the mathematical expression of heat flow. In any case, an understanding of the thermal aspects of welding, however detailed, is of interest from the metallurgical point of view only if we know how to take into account the influence of the factors in which the heat source supplies the variations vis-a-vis the phenomena we are studying.

The thermal study of welding also gives access to the study of tenso-thermal phenomena, because these, like metallurgical phenomena, depend on data established by the thermal study. The computer analysis approach has been responsible for much progress in our understanding of deformation and stress fields generated by welding.

Finally, the development of automation and robotics applied to welding has highlighted a third point of interest in the thermal study of welding. It has in effect become possible to 'pilot' an operation, i.e. to maintain control at fixed values by monitoring temperature evolution at a given point during the operation or the temperature difference between two points, by using appropriate means.

Purpose of the study

Knowledge of the temperature reached is a necessary condition for the metallurgical interpretation of all phenomena resulting from a thermal effect. In addition to this temperature condition, there are one or more time conditions, concerning heating, soaking or cooling. To interpret phenomena arising during the welding operation at a given point of the assembly, it is necessary to know the thermal cycle of welding, i.e. the temperature variation θ as a function of time t. The corresponding curve $\theta = f(t)$, plotted at point A close to a weld (Fig. 2.1) gives us information we need about the magnitude of variables, namely:

– Maximum temperature reached θ_m
– Soaking time T_s above a temperature θ_s
– The cooling law, translated by cooling time $T_{R(\theta_1\theta_2)}$ between two temperatures θ_1 and θ_2 or by cooling speed V_R at a temperature θ_R.

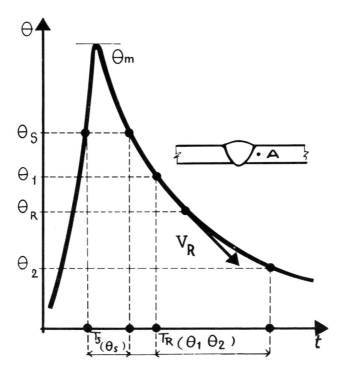

2.1 Parameters relative to the welding heat cycle $(\theta = f(t))$ at point A close to a weld: θ maximum temperature reached; $T_s(\theta_s)$, time above temperature θ_s; $T_R(\theta_1 \theta_2)$, cooling time between two temperatures $\theta_1 \theta_2$ (generally 800 and 500 °C in steel); V_R speed of cooling at temperature θ_R (slope of the tangent to the cooling curve).

Thus determined at a point in the assembly, the magnitude of variables associated with the heat cycle gives us access to the knowledge or anticipation of metallurgical phenomena arising at that point, such as structural modifications or transformations on heating and/or cooling. But if we wish to have access to the topographical distribution of those modifications around the area of the weld, we need to know the curves which translate the heat distribution and, in particular, curve $\theta_m = f(x)$ (Fig. 2.2) giving the variation of the maximum temperature θ_m reached at each point as a function of distance x. As each phenomenon, as a first condition, is characterised by the necessary θ temperature, we can, due to curve $\theta_m = (fx)$, site the exterior limit of the zone where that phenomenon is liable to occur. Thus we now have a quite general definition of the 'heat affected zone' or HAZ where this phenomenon takes place. For steel welds, this expression is reserved for the zone 'austenitised' by the welding heat cycle, i.e. the zone which is limited by the isotherms corresponding to transformation points A_3 (complete austenitisation) and A_1 (partial austenitisation).

In conclusion, if we limit our study to the metallurgical effects of welding, we have to establish on the one hand the welding heat cycles at each point near to the weld, in order to know the nature of the phenomena these welding cycles generate, and on the other hand the heat distribution, to interpret the extent and topography of these phenomena and their consequences.

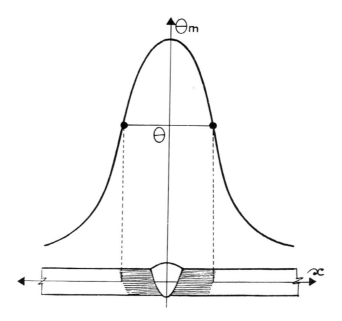

2.2 Thermal distribution $(\theta_m = f(x))$ on either side of a weld and definition of the heat affected zone by a metallurgical phenomenon occurring on heating to a temperature higher than θ.

As to the means employed to achieve these two objectives, they are of two kinds, one experimental and the other mathematical. Indeed, a synthesis of the two approaches is currently used with the help of a computer.

The experimental approach of the thermal aspect of welding is carried out by means of thermocouples. If the product is thick, the hot weld is inserted in a hole drilled in the part to be measured, distance x in relation to the fusion boundary being measured after sectioning (Fig. 2.3a). Measurement requires certain precautions relative to the thermocouple itself (which must be of the smallest possible diameter), the hot weld being welded integrally to the part at the bottom of the hole, and to the interference due to the actual welding, for some welding processes. A large degree of simplification was provided by the systematic use of implants*. These were small cylinders of metal of the same heat conductivity as the metal used in the experiment, prepared in advance and fitted with a thermocouple (Fig. 2.3b), then inserted with gentle friction into a hole drilled in the part being tested, so that the hot weld was at depth p in relation to the surface on which the experiment would be carried out. For a weld bead laid on that surface, at that distance, one achieves the same curve $\theta = f(t)$, whether the thermocouple is lodged in the part or in the implant. This similarity led to an important development, which we shall find again in studying cold cracking in steel welding (see Chapters 7 and 8).

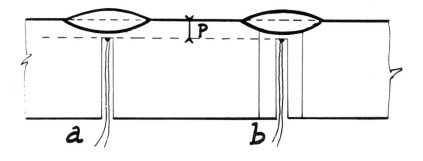

2.3 Recording curves $\theta = f(t)$ by means of electric thermocouples: a) Couple inserted in a hole drilled in the component; b) Couple lodged in a previously prepared implant inserted into the component. If depth p is the same, arrangements a and b supply the same recording θ f(t) for the same weld bead.

From the measurement thus made, (of necessity a pinpoint measurement), we obtained the magnitude of variables listed above concerning temperature and time which characterise the welding heat cycle. We can go further, either by instrumenting the thermo-electrical circuit, or by processing the curves themselves with the aid of a computer, to arrive by heat analysis at the transformation diagrams for steel under the effect of actual welding operations. Another technique requires the use of simulated cycles.

---

* H GRANJON, R GAILLARD: Possibilities offered by small specimens for studying hardening and underbead structure of steel welds. Soudage et Techniques Connexes - No 9/10, 1964, 343-45

Knowing the pinpoint heat cycles, it is only possible to obtain access to the heat distribution, and in particular to curves $\theta_m = f(x)$ by taking from each cycle recorded the maximum temperature reached. With welding of thin products, however, there is the possibility of an overall display of the heat distribution, at least on the reverse side of the parts. This possibility is offered by the use of coatings applied to the reverse side before welding. These coatings, called reference colours,[3] are characterised by a change of colour at fixed temperature; the change is maintained after cooling. We can thus observe the development of isotherms, defined later, and gain access straight away to the curve $\theta_m = f(x)$, at least in the range of temperatures which the reference colours withstand without destruction. Finally, recent progress in thermography has already enabled some researchers, using this technique, to tackle a welding heat study and even implementation of robotic welding.

For a long time the mathematical approach remained the prerogative of a few specialists, because of the difficulties inherent in applying heat conductivity equations to welding. We can nevertheless, while we are on the subject, cite the names of Rosenthal* and Rykaline** whose works are still authoritative. The difficulties involved in this type of calculation lie in the heat source, and in the metal being welded.

With regard to the heat source, the first problem is in its movement and, even if it is immobile, in the speed of its intervention. We are therefore a long way from the stationary condition, the situation which is assessed in standard heat flow calculations. Nevertheless, we shall see later that, for processes using a moving heat source, it is possible to reason in the 'quasi-stationary' condition, which we shall define.

Also, in the interests of simplification, we are obliged to accept that the source is pinpoint, which, depending on the processes concerned, is far from being the case. Ultimately, we have to involve the heat source yield, if it is known.

With regard to the metal being welded, the calculation should, along the entire length of the temperature-time curve, take account of the variation of the heat conductivity and calorific capacity characteristics, as modified by the transformations, as is the case for steel. Finally, we must obviously take account of the heat flow conditions as affected by thickness which led those who did the calculations to propose solutions relative to thin products and to very thick products, even though able to suggest solutions in between these two extremes. Moreover we shall see that the notion of thickness is not independent of welding process and procedures.

Nowadays, the use of computers makes it possible to refine calculations and introduce experimental results which extend validity. It is the results of these calcula-

---

* D ROSENTHAL: Mathematical theory of heat distribution during welding and cutting
** N N RYKALINE: Calculation of welding heat processes. Soudage et Techniques Connexes No. 1/2, 1961, pp 5–38

tions which are used to program simulations permitting metallurgical study of welding effects or leading to models relative to the distribution of deformations and stress.

General characteristics of the heat cycle and heat distribution

Quasi-stationary condition

If, using thermocouples, we measure curves $\theta = f(t)$ at points $A_1, A_2 \ldots$, etc, situated at the same distance from an arc weld bead and at increasing distances from the origin of that bead 0, presumed to have started in the middle of the metal plate, we obtain the arrangement described in Fig. 2.4 in relation to the start of the weld bead 0 for each experimental run; this lead us to the following comments:

a) The maximum temperature reached θ_m increases with the distance covered from the beginning of the bead, then stabilises at a value which becomes constant, whilst the heat source moves at an even speed, energy remaining constant.

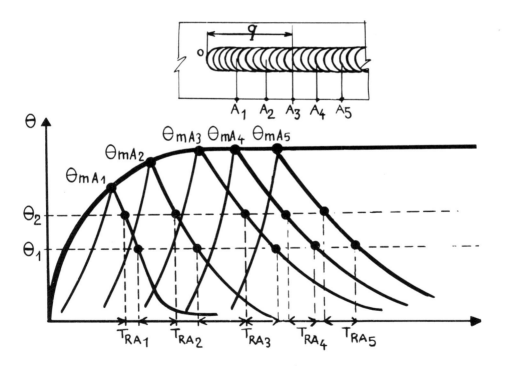

2.4 Evolution of curves $\theta = f(t)$ at a given distance from a weld bead in terms of the distance from initiation point 0. The quasi-stationary condition, which is established for distance q, is characterised by the identity of the temperature/time curves, resulting from the inequality of maximum temperatures, of time at high temperature and cooling time. First, the maximum temperatures increase, together with cooling times.

b) At the same time, cooling laws, measured by cooling times between two temperatures, become identical, as are the soaking times (not illustrated on the figure). In other words, curves $\theta = f(t)$ become superimposable as soon as the heat source has cleared a certain distance q.[4]

The arrangement thus described takes account of the quasi-stationary condition, which results from the fact that from a certain distance, covered in a steady movement by the heat source, the energy expended by conductivity in the part is continually compensated for by the energy supplied at the heat source. This can also be expressed by saying that, in relation to mobile heat source S (Fig. 2.5) beyond distance q, the isotherms remain the same as they were and move with the source. Their limits are lines parallel to the bead as shown by the reference colours. We shall later find an interpretation of this arrangement, calling on the notion of 'heat solid'.

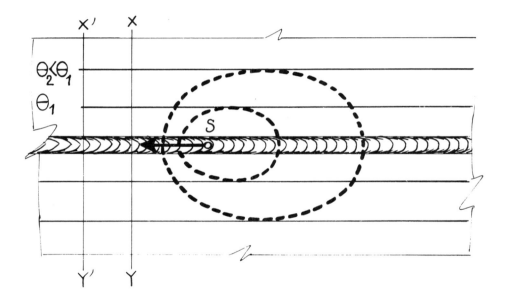

2.5 Position of the isotherms (broken lines) at the moment when the heat source passes point S. The movement of these isotherms in the quasi-stationary condition generates lines (un-broken) parallel to the weld bead, of equal maximum achieved temperature. A metallographic section perpendicular to the bead supplies the same information regardless of its XY or X'Y' position.

As indicated above, the standard considerations concerning the thermal study of welds and their consequences were drawn up in the quasi-stationary condition, and the formulae supplied in the literature are as a rule valid only in that condition. We should not however neglect the situations excluded in that condition, such as the beginnings and ends of beads, short welds, tack welds, fixings of various fittings, etc, which involve cooling speeds significantly higher than in the quasi-stationary condition. We shall return to these situations during the study of hardening in steel welding (Chapter 8).

Finally, of course, none of this applies to welding processes not involving movement of the heat source.

From the metallurgical examination point of view, the existence of the quasi-stationary condition is important because it justifies examination of welds on the transverse sections XY or X'Y' perpendicular to the welding line; the position of these sections is of little importance in so far as samples are taken from the part where the quasi-stationary condition was effective when the weld being examined was made.

Welding heat cycle: temperature/time curves

Because of the existence of the quasi-stationary condition, we can limit our determination of temperature-time curves to points A, B, C ... distributed on a straight line XY perpendicular to the weld line, traced from heat source position S. Whatever this position in the quasi-stationary condition zone, the results will be the same, shown as a diagram in Fig. 2.6 which calls for the following comments:

a) The temperature-time curves are situated one above the other as the distance between the measuring point and the weld line increases. In particular, maximum temperatures reaching θ_m decrease more or less abruptly depending on the temperature gradient characterising the process and procedures being studied. Naturally, all the curves unite asymptotically at the level of the initial temperature of the parent metal.[5] These two comments, which are obvious, have consequences which are less evident.

b) The maximum temperatures reached θ_{mA}, θ_{mB}, etc, which decrease from the weld line, are passed for times Tm_A, Tm_B, etc, which increase; this difference results in curve $\theta_m = \varphi(t)$ which separates the range of increasing temperatures from the range of decreasing temperatures. In practice, this expresses the fact that in the vicinity of the heat sources, the closest points are already in the process of cooling from their maximum temperature whilst the farthest points are still in the heating phase. We can understand the point of this observation if we think about the phenomena which are controlled by temperature variation (transformation, expansion, shrinkage) and whose timing is thus explained. In particular, when we examine the transverse section of a weld, we see the metallurgical aspects which are characteristic of those phenomena, but the structures that we observe have not appeared at the same time.

c) Given the arrangement of the curves, cooling speed measured from maximum temperature decreases when the distance from the weld line increases. But the cooling curves rapidly constitute quite a tight belt and we see, for example at points A and C, T_R times practically identical between temperatures θ_1 and θ_2 in that belt. So we can speak of a cooling speed or time associated with a welding condition. In

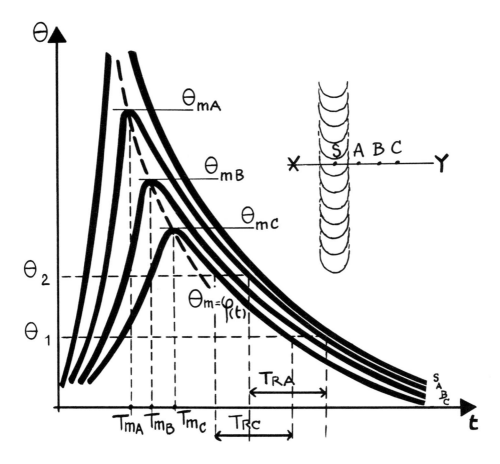

2.6 Group of temperature/time curves obtained at different measurement points along a line perpendicular to the weld line during the passage of heat source S. Note the displacement in the times of maximum temperatures and the quasi-identity of cooling times in the lower part of the curves.

particular, to study the transformation of steel on cooling, we most often choose as reference the cooling time between 800 and 500 °C, a particularly critical temperature gap in this case. However, we sometimes refer to the total cooling time, for example when we wish to take account of the influence of hydrogen diffusion. To make this clearer, the order of magnitude T_R (800 – 500) involved in steel fusion welding, varies from a few seconds to a few tenths of seconds depending on the welding processes and adjustments adopted.

Heat distribution: temperature/distance curves, heat solid

To describe the spatial distribution of welding heat cycles, Portevin and Seferian[*] proposed, under the name 'heat solid' (Fig. 2.7) a

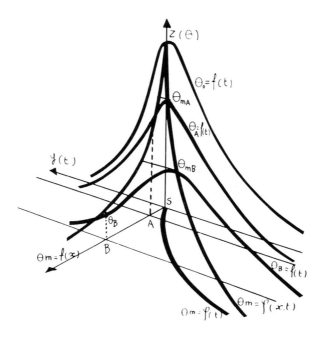

2.7 Portevin and Seferian heat solid: Distribution of instantaneous temperatures around the heat source at the moment of its passage through S during its movement along centreline Sy (times and distances covered). Temperatures are plotted parallel to centreline Sz for each distance x in relation to the weld line.

representation which takes account in three axes (Sx distance to weld line, Sy distance along that line, i.e. time axis, and Sz on a perpendicular to plane Sxy along the length of which temperatures are traced) of the instantaneous condition of the temperatures reached around heat source S at the moment of its passage at the origin of the coordinates.[6] Because of the quasi-stationary condition, the heat solid remains the same as it was throughout the length of the weld line along which it moves, generating at each point the temperature/time curves we have just described. On the heat solid in Fig. 2.7, of which only half is shown, curves $\theta = f(t)$ at the points defined by their distance x to the weld line are sections by planes parallel to plane yoz. On these curves, the maximum temperatures are different in time, as taken account of in the lefthand curve $\theta_m = \varphi(x,t)$ which is shown on the solid, whilst curve $\theta_m = \varphi(t)$, the importance of which we have already emphasised, represents only a projection of those maximum temperatures on plane xoz.

If we now cut the heat solid into planes parallel to plane xoy and project the intersections on to plane xoy, we obtain, like the contour lines on a topographical map,

* A PORTEVIN, D SEFERIAN: Heat study of torch and arc welding.
XIth International Congress of Acetylene and Autogenous Welding – Rome, June 1934, vol. III, pp 269-97

instantaneous equal temperature curves (not shown in Fig. 2.7); movement of these curves produces isotherms, as we have already shown (Fig. 2.8), lines parallel to the weld line itself. Each isotherm corresponds to a maximum temperature for which the time difference in relation to the position of source S is, as suggested above, indicated by curve $\theta_m = \varphi(t)$, itself a projection of the lefthand curve $\theta_m = \varphi(x,t)$ marked on the heat solid.

So in a single representation there is the data for temperature, time and distance necessary to understand the phenomena generated by welding. Depending on the process involved, the heat solid is more or less sharply pointed, but all welding processes with movement of the heat source depend upon this representation.

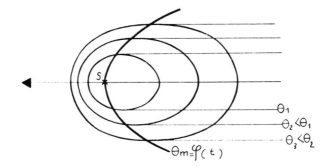

2.8 Formation of isotherms in plane xsy by projection of curves of equal temperature from the heat solid and movement of that solid.

Welds without movement of heat source

When a weld is made without movement of the heat source,[7] there is no quasi-stationary condition because the weld is achieved by an input of localised and momentary energy sufficiently intense not to be compensated for by heat flow in the assembled parts, at least whilst the process of achieving metallic continuity is in progress. We can, however, as for the other processes, interpret the metallurgical phenomena involved by considering the temperature-time cycles $\theta = f(t)$ and the distribution of resultant structures and properties by means of curves $\theta_m = f(x)$ showing the distribution of the maximum temperatures reached during the operation.

By and large, these are relatively quick processes, on account of the interruption in energy input which accompanies completion of the welding and also, for some processes, on account of cooling by conductivity outside the part itself, for example in the jaws or the electrodes of resistance welding machines.

Factors of the cycle and heat distribution

Influence of the parent metal

The parent metal influences the welding heat process by its own heat characteristics, the mass and arrangement of the parts to be assembled and finally by the initial temperature of those parts.

a) The nature of the parent metal is involved by its heat conductivity and capacity characteristics which act in the same sense, namely that their increase increases the temperature gradient and, consequently, reduces soaking at high temperature and increases cooling speed. This is the case for stainless steel compared with ordinary steel. This is also the explanation of the operational difficulty in arc welding copper, resolved by preheating which, here, has the effect of reducing the temperature gradient and thus making it possible to achieve a molten pool. We can also mention spot welding of aluminium which, compared with steel, requires more energy delivered in a shorter time, as is also the case for stainless steel. Finally, we must note (we shall deal with this in more detail later) that calorific capacity is involved in the calculations only for thin parts and not for large parts.

b) With regard to the mass of parts, it should be noted that this is a relative notion which cannot be separated from the influence of welding conditions since, for example in steel arc welding with coated electrodes, a whole range of energy settings can be used to deposit a weld bead on the surface of a metal plate of a given thickness. So if we trace a graph (Fig. 2.9) giving the cooling time T_R in terms of thickness (e), at the weld bead for two different welding energies E_1 and E_2, we see that cooling time stabilises from a certain thickness, called limit thickness. The greater the quantity of energy, the thicker this limit becomes. So, a product is thick, from a welding point of view, if its thickness is greater than the limit thickness associated with the energy used. In this case, the isotherms are distributed symmetrically in the mass in relation to the weld line on which they are centred, taking a cylindrical shape (Fig. 2.10c). It is this arrangement that Rykaline's formulation refers to; this is discussed later. On the other hand, with a product described as thin – or more precisely a product affected throughout its entire thickness by fusion – the isotherms are distributed symmetrically in relation to the symmetry plane of the weld line itself (Fig. 2.10a) and, as we shall see later, the formulae are different. The situation for parts of average thickness (for which cooling time falls within the descending part of the curves in Fig. 2.9) is illustrated by Fig. 2.10b which shows isotherms deformed by the effects of cooling by radiation on the lower surface of the part.

The relative aspect of this notion of thickness results in an apparently paradoxical consequence. Each time a welding process permits butt assembly of two parts where their entire thickness is affected by fusion (of course in a single run) we have a symmetrical distribution of isotherms, i.e. as shown in Fig. 2.10a. This is particularly the case for electron beam and electroslag welding, to which the formulations relative to thin products are applicable, even with very thick products.

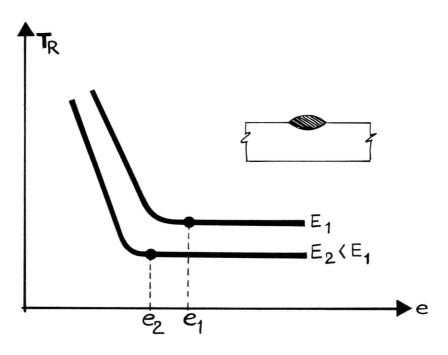

2.9 Variation of cooling time depending on thickness, for a bead deposited by arc welding. Limit thickness is associated with welding energy.

c) To take full account of the role played by the mass of welded metal vis-a-vis the welding heat effect, we must, in addition to thickness, take account of the arrangement, or, assembly geometry, of the parts. Indeed, with equal energy and thickness, a weld bead may for example be deposited on the surface (a), or be used to make the first run on a single V butt weld (b), or to constitute a lap weld called 'bithermal' (c) or 'trithermal' (d) depending on the proximity of the bottom edge in relation to the weld (Fig. 2.11). Each of these cases corresponds to a heat flow, therefore to a different heat cycle, even though resulting from the same energy input.

The recommendations concerning welding of steels take account of this by coefficients affecting the energy used according to the geometry of joints for associated thicknesses (see Chapter 8).

d) The initial temperature of the parent metal plays an important role vis-a-vis the welding heat cycle and distribution, since the temperature gradient depends directly on them. If, all other things being equal, we pass from θ_o to θ_p for the initial temperature of the parent metal, we get the temperature/time curves illustrated in Fig. 2.2a and distance-maximum temperature curves of the type shown in Fig. 2.12b in the proximity of a butt weld. On curve $\theta = f(t)$ we see that at the price of an increase in maximum temperature reached and in soaking at high temperature, the

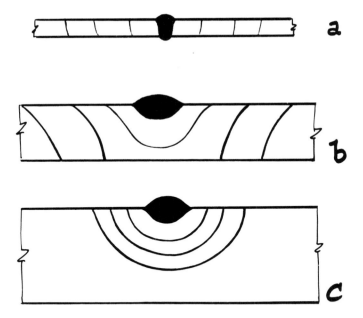

2.10 Arrangement of isotherms through the thickness of the welded product: a) Product welded over its entire thickness: isotherms symmetrical in relation to joint plane; b) Deposit on product of average thickness: isotherms symmetrical in relation to plane perpendicular to surface but curves in towards interior surface; c) Deposit on product thicker than limit thickness: cylindrical isotherms centred on centreline of deposit.

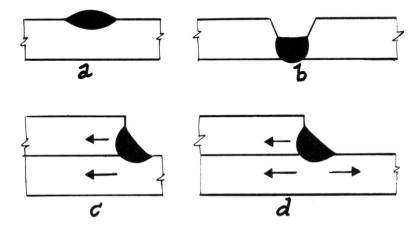

2.11 Examples of the influence of joint geometry on cooling time, for a weld bead with identical energy: compared with bead a, less rapid cooling time for bead b, more rapid for bead c (bithermal), even more rapid for bead d (trithermal).

increased initial temperature results in a slow-down of cooling: cooling time is increased, whether it is the total time or the time between two temperatures. Such is the effect of preheating, which may be global if it affects the whole part, or local if it is applied only to a limited area around the joint. Another (unlooked for) effect is apparent on the curves $\theta_m = f(x)$ where preheating increases the extent of the heat affected zone, as it decreases the temperature gradient.

Lowering the initial temperature results in opposite effects, such as acceleration of cooling. Hence the precautions taken when welding steel during cold weather. We sometimes use the word 'preheating' to describe the preheating limited to the beginning of welding operations and carried out at a moderate temperature, in particular to heat parts which have been stored outdoors and recently brought into the workshop for welding.

With regard to preheating, it would be a good idea to make a few observations which are of great importance in steel welding, as detailed in the following paragraphs.

Since one of the effects of preheating is to increase the maximum temperature achieved during welding, hence an accentuation of overheating of the parent metal, it is necessary to keep to the minimum preheat temperature compatible with the objective sought. This is particularly valid for relatively small parts where excessive preheat can be damaging because of the over-excessive temperature those parts can reach by the end of welding.

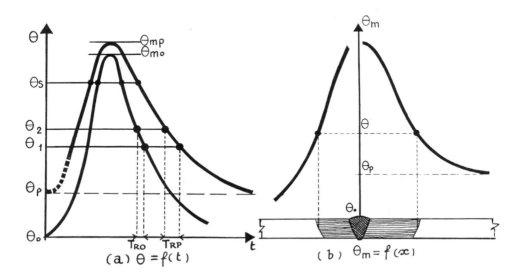

2.12 Influence of initial temperature: a) On heat cycle: preheating to θ_p increases maximum temperature and time at high temperature. It also increases cooling time. b) On thermal distribution: preheating enlarges the heat affected zone by a phenomenon occurring at temperature θ.

If preheating is global and the part to be welded relatively large, the energy input at the moment of welding is sufficient to maintain the desired temperature, even if welding operations are prolonged. But it is nonetheless necessary to take care that such is indeed the case. This applies particularly to the run-off test plates used for the qualification of welding procedures or manufacturing inspection. Care should be taken to ensure that these test pieces are truly representative, from the point of view of mass, of the parts they are supposed to represent. If welding is done in several runs, even more care must be taken on this point as temperature increases spontaneously from one run to the next, as we shall see later.

If preheating is local, i.e. limited to the joint and its immediate surroundings, there is a temperature gradient between the zone thus preheated and the remainder of the part. So, if the part is large, it is necessary to keep the heat source used for preheating in action, so as to compensate loss by conductivity at the joint. In the same vein, local preheating should never be prescribed solely in terms of temperature; for example, it is wise to indicate the minimum distance in relation to the joint at which the required temperature must be achieved, or indeed to fix the time during which the temperature must be maintained at the joint.

In the case of local preheating, it should not be forgotten that the effects of expansion and shrinkage can thwart the benefits of preheating because of the creation of stresses. So it is, for example, for the arrangement illustrated in Fig. 2.13, which corresponds to a crack test, called H test. The local preheat of the shaded zone around weld S is, by expansion of that zone, causing the joint edges to draw closer together. After completion of the weld, the prevented shrinkage of the heated zone causes higher stress than if the welding had been done without preheating. So account must be taken of constructional arrangements around welds to decide if and how local preheating can be applied.

Finally, coming back to overall preheating, if this treatment is applied to a very large part, cooling of that part after welding may be long, such that the effect

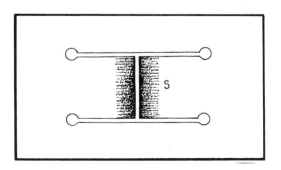

2.13 Example of arrangement for which local preheating is contra-indicated: the weld is put under traction during cooling.

produced is that of postheating, which enables the preheating temperature to be significantly reduced. In effect, postheating is an operation which consists of interrupting post-welding cooling, before complete cooling has occurred, in such a way as to maintain the welded joint and its immediate surroundings, or even the whole of the welded part, at a preselected temperature, then to allow cooling to continue normally. Postheating almost always takes place at the same temperature as preheating (Fig. 2.14) and combines its effects with those of preheating, but this is not obligatory and postheating can take place at a temperature higher or lower than that of preheating, or even without any preheating at all, for example in spot welding. The effects of postheating are studied in Chapter 8.

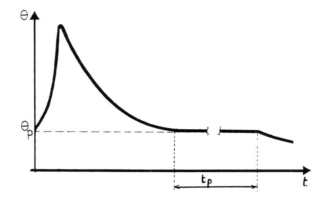

2.14 Principle of postheating where it is practised at a temperature equal to that of the preheating: θ_s temperature; t_p duration of postheating.

Influence of welding conditions: energy, environment

The influence of welding conditions can be expressed in terms of the energy used during the operation. For some processes such as gas welding, plasma welding and, to a lesser degree, electron beam welding, the energy range necessary to carry out a weld on a given thickness is relatively narrow. On the other hand, for arc welding processes, the energy input can vary widely, in such a way that energy is an important factor of the heat cycle and heat distribution.

In arc welding, the energy input may be evaluated by calculating the value of the expression $E = \dfrac{UI}{V}$, where U (in volts) is the arc current, I (in amps) is the intensity of the welding current and V (in cm/min) is the welding speed. For a given intensity setting, we can evaluate energy by measuring the time t taken by the arc to cover a length L. If L is expressed in cm and t in seconds, the nominal energy is written as $E = \dfrac{UI}{L}t$ and expressed in joules/cm.[10] To explain more clearly, we can say that in arc welding with coated electrodes, the energy varies from a few thousandths to a few tens of thousandths of J/cm, whilst with submerged-arc welding, energies of the order of 100th of a thousand J/cm are involved.

In reality, the energy involved, thus calculated, is not the energy effectively transmitted to the part, because each heat source is characterised by a coefficient characterising its efficiency. For example, with submerged-arc welding, this is 80 to 95%, whilst it is only 70 to 80% in welding with coated electrodes. As this efficiency is assumed to be constant for the various energy settings, we can only associate data relative to the heat cycle and its consequences (e.g. underbead hardness) to nominal energy for a single welding process. In the range of energy common to two processes, the effects may be different.

Be that as it may, as we have already seen with regard to the influence of thickness, increasing the energy used extends the heat affected zone, increases the soaking at high temperature and slows down cooling (Fig. 2.9). The result is that processes or welding procedures using high energy are characterised by a tendency to overheat (associated with the soaking time) and, in steels, a lesser tendency to hardening (associated with cooling speed).

Finally, at equal thickness, we have two variables to act on the welding heat cycle, namely initial temperature (preheat or not) and nominal energy. Depending on the circumstances (workshop or site, work in a confined space or in the open air, welding position, etc) it is possible to manipulate one or other, or both.

In addition to the standard factors which we have mentioned, others may be involved in the context of what we call here the welding environment. These are listed below.

In welding thin products, in particular with automatic welding, the parts to be welded are fixed by jigs. These jigs absorb heat and thereby modify the theoretical heat conditions of welding.

Climatic conditions also play a role, by the effects of temperature (already discussed) and also wind and damp. Preheating, as already mentioned, helps to avoid condensation on parts to be welded.

The recent development of underwater welding, in particular hyperbaric, has shown the importance of the environmental factors of pressure and the atmosphere surrounding the electric arc. Increased pressure causes constriction of the arc and consequently an increase in its temperature[9] and the use of a heat conducting gas such as helium makes the losses by convection significant. This is not taken account of in traditional welding. These two causes give rise to a single effect, namely accelerated cooling, at equal nominal energy.

Summing up mathematical formulation

The combined influence of the nature of the metal, mass and initial temperature of the parts, and finally of the energy used, has resulted in formulations from which those of Rykaline (already mentioned) are selected with regard to the soaking time at high temperature and the law of cooling.

For a thick product (in the sense we have already indicated) on which a weld bead is deposited, soaking time t_s cooling speed V_θ are written as:

$$(1)\ t_s = k_1 \frac{E}{\lambda(\theta_m - \theta_0)} \qquad (2)\ V_\theta = \frac{2\pi\lambda(\theta - \theta_0)^2}{E}$$

For single run butt welding, or depositing a bead on a thin metal plate, these two variables are written as

$$(3)\ t_s = k_2 \frac{E^2}{\lambda C_\gamma (\theta_m - \theta_0)^2 e^2} \qquad (4)\ V_\theta = \frac{2\pi\lambda C_\gamma (\theta_m - \theta_0)^3 e^2}{E^2}$$

In these formulae:
λ is heat conductivity.
C_γ is volumetric heat capacity.
E is the energy used (assigned nominal energy of the yield coefficient).
θ_0 is initial temperature.
θ is the temperature at which cooling speed is calculated.
θ_m is the maximum temperature reached at the point in question.
k_1, k_2 are the coefficients as a function of temperature θ to which the calculation applies.

We shall not go into detail on the solutions proposed for single run welding of products of intermediate thicknesses; we shall merely comment on the above formulae, which quantitatively explain the influence factors which we have studied in a qualitative manner earlier in this book.

With regard to the nature of metal, we see heat conductivity in both types of formulae, whilst volumetric heat capacity C_γ is involved only for thin products. This results from the difference already emphasised between thin and thick products, from the point of view of the arrangement of isotherms and the conditions of the quasi-stationary state. Moreover, if that was not the case, it would be impossible to deposit a weld bead on thick metal.

As for the influence of the mass of parts and their arrangement, suffice it to recall that a coefficient taking account of joint geometry must be assigned to the energy used.

The influence of initial temperature θ_0 is taken into account by its presence in the denominator for the soaking periods at high temperature and the numerator for cooling speeds. In both cases, this influence is accentuated for thick products, as shown by the squaring of the soaking period and cubing for thin products. As already stated thin products require less preheating than thick products and excessive preheating encourages overheating.

Similarly, energy, which is involved in the numerator for soaking times and in the denominator for cooling speeds, has more influence on thin products than thick products, because it is squared for thin products. That is the reason why Rykaline proposed the allocation of the coefficient 3/2 to energy for welding the first run on chamfered metal of average thickness.

For the rest of this chapter, these formulae will not be used, but it was necessary to mention them to enable the reader to have a better understanding of recommendations made concerning steel welding.

Welding in several runs or cycles

The data given thus far concerns the heat aspect of fusion welding in a single run (i.e. with a single pass of the heat source) or indeed on the first run of a weld done in several runs. It would be advisable now to examine temperature evolution during subsequent runs. This evolution depends essentially on:

- initial temperature,
- the number and arrangement of runs and the conditions in which they are applied,
- the time lapse between runs,
- the position in relation to the weld of the point at which temperature variation is monitored.

Despite the variety of possible situations in terms of these factors, we can, for example, describe the heat effects of multipass welding (Fig. 2.15a) in the case of a V butt weld, made in three runs without preheating, for which the temperature-time curve is determined at point A affected from the first run.

This first run results in an initial temperature rise to maximum value θ_{mI}, followed by cooling which, characterised by speed V_1 until temperature θ_v, continues until the initial temperature if the second run is carried out after some delay. This is not generally the case, and the second run is applied before the temperature at point A has returned to initial temperature θ_0; it is still at $\theta_1 > \theta_0$ (10) when a heat cycle commences because of the second run, this cycle being characterised by a maximum temperature θ_{mII} and cooling speed V_{II} at temperature θ_v. But this cycle differs from the first one: θ_{mII} is less than θ_{mI} because the distance of measurement point A in relation to the second run is greater; also V_{II} is less than V_I for the same reason and also and especially because the heat cycle of the second run is influenced by an effect similar to that of preheating, because that cycle starts from $\theta_1 > \theta_0$. The same applies to the third pass, which, for the same reasons and for the arrangement

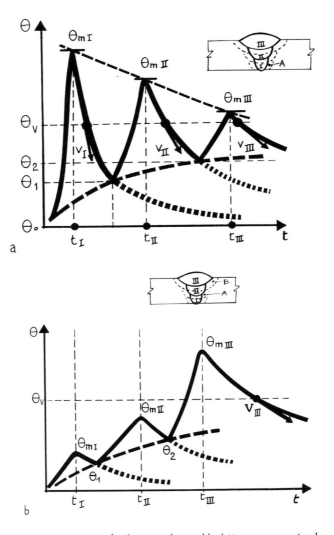

2.15 Example of heat cycles in the vicinity of a three-pass butt weld: a) Measurement point close to first pass: the maximum temperature θ_{mI} is the highest as is the cooling speed V_I. Both decrease during the subsequent passes, whilst the initial temperature increases for each pass from θ_0 (broken line curve). The maximum temperature reached may exceed by twice or even three times a given value, for example point A_3 of steel. The dotted curves represent the cooling which would follow each pass if the next pass was not made. Hence the importance of the time gap between passes; b) Measurement point close to the third pass: the maximum temperatures are increasing and may exceed a given value only once.

described, the maximum temperature θ_{mIII} and cooling speed V_{III} are respectively less than θ_{mII} and V_{II}.

If, for identical operative conditions, we now record curve $\theta = f(t)$ at point B (Fig. 2.15b) close to the third and final run, we would see that at this point the heat shock due to the first two runs is relatively weak; it is only on the third run that maximum temperature is reached at this point, whilst cooling speed is moderated by the preheating effect already described.[10]

Of course, the heat cycles thus described also depend on the run rate, i.e. the time lapse between them, during which temperature falls along the broken line curves in Fig. 2.15a and b.

Finally, the heat study of a multipass weld leads to the following observations:

a) It is in the vicinity of the first run that we see the severest heat cycles as regards maximum temperature reached and rapidity of cooling. Subsequent runs give rise to ever diminishing cycles because of the progressive rise of the initial temperature which increases after each run.

b) Depending on the arrangement of the weld runs, a single temperature θ (for example point A_3 for steels) may be exceeded two or more times in a given region during subsequent runs). The structure finally observed in this region is the result of the succession of the corresponding cycles.

This observation is also as valid for the heat affected zone as for the weld metal, which is also reheated and, where applicable, transformed during successive runs.

c) In a multipass weld, there are always regions corresponding to the last run (or runs) where the structure is a result only of the last heat cycle which produced it. However, that cycle is in principle less brutal than the preceding cycles, at any rate as regards cooling speed, especially as the last runs are most often made with higher energy input than earlier runs. This is the case, for example, with the last run of a pipeline weld which the Americans have christened 'hot pass'.

d) The important role played by the time lapse between successive runs vis-a-vis the temperature at the moment of each run explains the necessity of specifying this factor when stipulating the welding procedures. Usually, specifications stipulate the 'temperature between runs' i.e. the temperature guaranteed at all times by the chosen sequence.

The involvement of several heat cycles is not the exclusive prerogative of fusion welding; it is found also in resistance welding of steel, especially spot welding where, after an initial heat cycle (obtained by joule effect in the metal and where the parts are in contact) has caused the formation of the point, a second cycle may intervene to cause reheating (by joule effect only) of that point and its surroundings (Fig. 2.16). Here again, apart from the welding and reheating current intensities I_S and I_R, the important factor of the procedures is the dead time t_m which separates the two successive currents. This is a very short time possible only with electronic mechanisms.

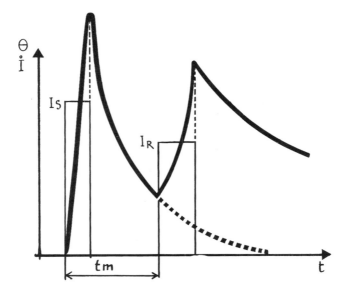

2.16 Heat cycle in resistance spot welding with use of a reheat current: I_s welding current, I_R reheat current (called 'annealing' in the case of steels). The efficiency of this reheating depends on the dead time t_m which separates the two passages of current.

Characteristics of the principal processes of welding and thermal cutting

Gas welding

Gas welding belongs to the category of processes for which energy – or flame power – needs to be regulated in accordance with the thickness to be welded, so that we can describe the heat cycle and heat distribution which are characteristic of this process, without the need to include the variables involved in arc welding. Also, contrary to what happens in arc welding, the molten pool only slightly exceeds the fusion temperature of the material welded.

We can summarise the heat regime of gas welding by saying that the heat solid[11] has a weak gradient, which shows a weak temperature gradient. The result of this is a slow heating, associated with a long soaking time and relatively slow cooling. Indeed, cooling speeds are expressed in degrees/minute, whilst in arc welding it is degrees/second.

From the metallurgical point of view, we can say that the soaking at high temperature encourages the formation of a coarse structure (overheating) which is also found in weld metal. On the other hand, the relative slowness of cooling generally excludes the phenomena characterised by a critical speed, such as hardening in steel.

Arc welding

Arc welding with coated electrodes can be studied at the same time as MIG or MAG welding, for these three processes cover the same energy range (approximately 5 to 30 $\times$ 10^3 J/cm) which corresponds to cooling times of between 800 and 500 °C of the order of a few seconds to 30 or 40 seconds depending on thicknesses and also initial temperature. In steel welding, understanding of the cooling times associated with energy, thickness and initial temperature makes it possible to foresee the transformations generated by the welding cycle, by means of diagrams (Fig. 2.17) drawn specially for welding.* Indeed, the rapidity of cooling and elevation of temperature achieved make standard diagrams unusable quantitatively. Methods of plotting diagrams relative to welding will be mentioned in Chapter 7. At the same time, as demonstrated by Fig. 2.17, we have

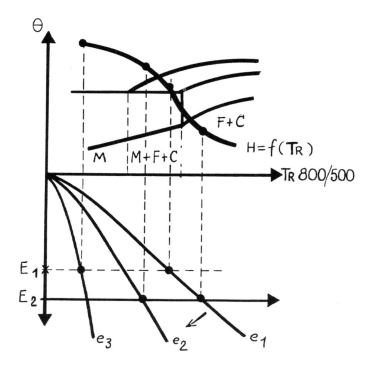

2.17 Correlation between energy E and thickness e, with the nature of the transformation on the one hand and with underbead hardness (bead deposited on the surface of a metal plate) on the other.
θ: cooling time temperature T_R, underbead hardness H, thickness e, energy E, martensite M, ferrite F, cementite C.

---

* IS/CETIM: Practical documentation on weldability of steels – 1976

access to forecasting 'maximum underbead hardness' by means of the curves accompanying the transformation diagrams. It turns out that, for many steel grades, the characteristic cooling times for arc welding, depending on settings, are higher or lower than those which develop hardening. Hence the importance of this correlation which is discussed further in Chapter 8.

When turning to submerged-arc welding, we should remember what has been said earlier, namely that if the coefficient expressing yield is not attributed to the nominal energy, we cannot trace the hardness/energy curves relative to the two processes, arc and submerged-arc welding, on the same diagram. We might be tempted to do so, since submerged-arc welding can be used with low energy inputs as with very high energy inputs, of the order of 100 thousandths of J/cm.

For such energy values (which are the most characteristic of the process) the extent of the heat affected zone and soaking period are significantly higher than in visible arc welding; cooling times are also longer. So we meet in the molten metal, as in the heat affected zones of steel welds, structures more affected by the soaking period at temperature (overheat), whilst the risk of hardening becomes less with the increase in energy. To reduce the effects of overheating, it is often preferred to operate with less energy but more runs, rather than a single high energy run, which often necessitates a subsequent heat treatment.

Another characteristic of submerged-arc welding is that, even for single run welding, the high energy which can be used may result in a molten pool whose volume and form create heat cycle dissimilarities depending on the position of zones explored in relation to the weld zone. This is particularly so if more than one filler wire is used. For example, if we plot curves $\theta = f(t)$ at two points A and B close to a single run submerged-arc weld (Fig. 2.18) we find two curves differing in cooling rate which seems retarded for point A, which results in a longer time at high temperature than for point B. This difference results from the shape of the molten pool (illustrated by a plan-view and longitudinal section), giving rise to a more prolonged soaking time for point A, which is affected by the heating earlier than point B. This means that a metallographical examination will reveal a zone with coarser grains at point A than in the region of point B. This zone is liable to appear during the side bend test as a failure which does not necessarily mean the presence of a defect.

Vertical electroslag welding

As indicated above, vertical electroslag welding (at least for butt assemblies) must, from the heat point of view, be dealt with on the basis of treating the assembled products as thin, regardless of their actual thickness. Indeed, these products are joined by a single run over the whole thickness, the filler metal being supplied by several wires or a consumable guide, in such a way that the isotherm surfaces in the heart of the parent metal arrange themselves more or less perpendicularly to the surfaces, and symmetrically in relation to the joint plane (Fig. 2.19). We will not waste time on the formulae provided by the literature,[12] but will simply note that, from the thermal point of view, this is a process which is slow

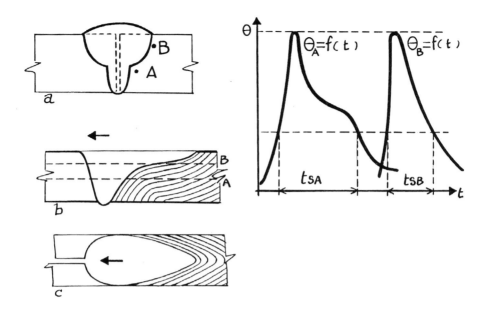

2.18 Heat cycles at two points A and B in the vicinity of a submerged-arc weld and interpretation on a transverse section (a), longitudinal section (b) and plan view of the molten pool.

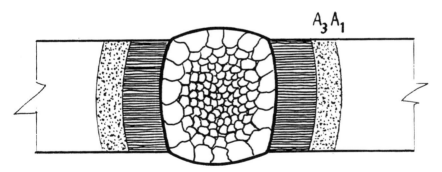

2.19 Arrangement of isotherms on either side of a vertical electroslag weld. Transverse section.

compared with the others.[13] The metal of the molten pool is maintained for a long time in the liquid state and solidifies slowly. In the heat affected zone, the time at high temperature is long and the consequently cooling is slow. These thermal characteristics mean that electroslag welding is a process which permits a real metallurgical preparation of the molten metal and, in particular, efficient degassing. Slow cooling produces no risk of hardening in steel welding, but the slow solidification of the molten metal and prolonged heating in the heat affected zone generate coarse structures which have to be eliminated by post-weld heat treatment or at least reduced by certain additions.

Electron beam and laser welding

As already mentioned, electron beam (EB) welding provides another example of a process permitting parts to be assembled throughout their thickness (Fig. 2.20a). Nevertheless, when there is not a sufficiently powerful gun available to obtain this result for the thickness in question, it is possible to make two runs, one on each side (Fig. 2.20b). In the first case, the formulation of the heat cycle corresponds to that of thin products, if the energy input is expressed in terms of unit area of the section of the welded zone.

As for the molten metal, its formation, generally fully provided by the parent metal (the use of filler metal is rare), constitutes a characteristic of the process. In effect, the electron beam perforates the base metal and creates a cylindrical hole whose walls are made up of the liquid metal[14] resulting from the fusion of the parent metal. It is the advancement of this liquid hole and the solidification behind it which provides the very narrow fusion zone, which characterises the process. The result, contrary to what occurs in other fusion welding processes (where energy is delivered at the point of impact of the heat source, leading to heat flow by conductivity), is that the electron beam delivers its energy over the total thickness which it affects, although at a loss, which can be taken as an operational criterion by measuring, in the case of Fig. 2.20a, the residual energy of the emergent beam. In case b, the beam gives out, resulting in a clear heat cycle difference, faster at the bottom point of the bead (i.e. at mid-thickness) than close to the impact point.

The process thus described implies a marked temperature gradient, therefore a sharp heat solid, hence brief heating, a narrow heat affected zone and rapid cooling, but the cooling time between 800 to 500 °C does not decrease with thickness as happens in arc welding. On the contrary, it increases by a few seconds for a thickness of between 10 and 20 mm, to around 20 seconds for 80 mm, which constitutes an interesting characteristic from the weldability point of view.

Given these characteristics, the principal preoccupation concerning EB welding lies in the performance of the weld metal (possibility of volatilisation of elements due to

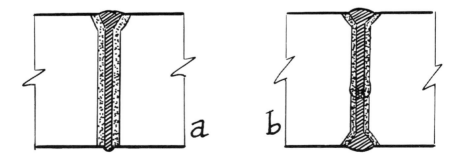

2.20 Transverse section of an electron beam weld: a) In a single pass over the entire thickness; b) In two passes, one each side.

the combined effect of vacuum and raised temperature) then rapid cooling and solidification. Another problem arises, namely that of the effect on the mechanical performance of the assembly of a fused and transformed zone, which is narrow and different in its properties from those of the parent metal.

From the thermal point of view, laser beam welding has something in common with EB welding as regards the concentration of energy involved, at least for thicknesses accessible to date to the beam ratings used. However, the heat transmission process is more closely related to that of arc or plasma welding than to that of EB welding because the transmission is effected by conductivity in the metal from the point of beam impact. Also, the energy balance of the operation must take account, depending on the metal welded, of the proportion of the laser beam energy lost by reflection on the molten pool surface.

Resistance spot welding

With spot welding, we enter the field of processes in which the notion of quasi-stationary condition plays no part. Heat distribution results from rapid local heating and cooling from the disappearance by conductivity of the instantaneous heat solid. The process is necessarily rapid, in order that the necessary temperature can be achieved by a heat input which momentarily exceeds the loss by conductivity in the part and the electrodes.

When carrying out spot welding, the duration of which is expressed in a number of periods in alternating current, the heating due to the Joule effect results successively from the contact resistance of the parts for assembly, then from the resistivity of those parts, which occurs as soon as a metal bridge begins to join them together. Incidentally, heating occurs on contact between electrode tips and parts; this is responsible for the mark which is called an indentation. When the weld has been obtained, at the moment when the current is cut off, the form of the temperature distribution curves throughout the thickness (Fig. 2.21) shows a big temperature

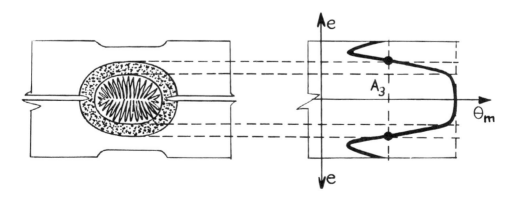

2.21 Distribution of maximum temperatures reached during execution of a resistance spot weld.

gradient, from the point itself, at fusion temperature, to the relative maximums reached on the surface due to the electrode/part contact resistance. Achieved at the moment of current cut-off, this heat distribution disappears quickly in the part and electrodes.

The temperature gradient results in a tight arrangement of isotherms around the point, so that those marking characteristic phenomena such as the transformation of carbon steel or the precipitation of carbon in stainless steels do not always emerge at the surface of parts.

As regards the rapid cooling, this renders steels capable of being hardened which, normally, is not the case with other welding processes. Hence the interest in the heat treatments we have already described in so far as they are compatible with acceptable sequences.

Thermal cutting

Thermal cutting consists of dividing the metal or alloy into sections by fusion by means of a mobile heat source to obtain localised fusion. To mention only two of the principal processes, the necessary heat energy can be supplied by combustion of the iron by means of an oxygen jet (oxygen cutting) in the case of 'black' steel (i.e. low alloy or non-alloy), or by direct use of a plasma jet. The weld metal is eliminated by the dynamic effect of the jet and there remains on the edges of the cut a zone affected by the heat cycle generated by the cutting operation.

In flame cutting, it is necessary to achieve and maintain the combustion initiation temperature along the entire length of cut; this is achieved by means of a 'heating

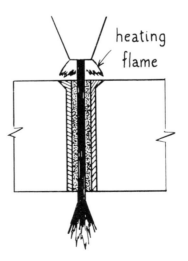

2.22 Constituent zones of a flame cut. It is possible to distinguish a fused zone (dotted area) maintained by combustion and a heat affected zone (oblique lined area).

flame' whose heat effect combines with that of the combustion to give the heat affected zone a particular shape (Fig. 2.22) which is not found in the vicinity of plasma cuts. As for electroslag or EB welding, cooling in the affected zone is slower in proportion to increased thickness: cooling times between 800 and 500 °C for steel go from a few seconds to about 20 seconds for thicknesses of 90 to 200 mm. As for the heat affected zone, this is the seat of a high temperature and cooling speed gradient, so that it is narrow vis-a-vis the thickness cut. The result is that when thermal cutting is used to prepare joints for welding, little attention is paid to its heat effect because the affected zone is usually covered by that of the subsequent weld. On the other hand, the use of parts with rough cut edges can pose problems, in particular fatigue problems, especially in the case of defective cuts containing faults liable to act as notches.

Finally, like welding, and for the same reason, flame cutting generates residual stresses and deformation. The crack phenomena which can result are rare, but cutting programmes must take account of deformation caused by the cutting operation so as to compensate for them as far as possible.

Deformations and stresses generated by welding

Mechanisms

This section concerning deformations and stresses, will be limited to a qualitative account which will develop as necessary what has been indicated earlier, namely that the metallurgical phenomena generated by welding[15] (when the process used involves a heat source) occur and manifest their consequences in a deformation and stress field which must be taken account of.

First of all, a few simple but standard figures will help the reader to understand how deformation and stress arise, the stress originating from the hindrance of deformation. In the following experiments,[16] we shall use prismatic bars, in steel for example, to heat them in their medium or to obtain fusion. The various conditions in which the bars are treated are illustrated in Fig. 2.23.

a) Free bar, locally heated evenly over a straight section (for example by induction) then freely cooled. This bar increases overall by a length of $2\Delta 1$ dependent on the width heated and temperature achieved, then returns to its initial length by cooling. There is no deformation or residual stress.

b) Bar treated as in a), but held at its ends between two fixed walls (without appreciable initial compression). During heating, the treated zone tends to lengthen as above, but is prevented from doing so because the bar ends are butting against the walls. The bar is then subject to compression, which causes plastic deformation of the heated zone; this deformation is encouraged by the lowering of the elastic limit resulting from the rise in temperature. In other words, it is the heated zone which compensates length $2\Delta 1$ of the above experiment by its plastic deformation (a bulge appears).

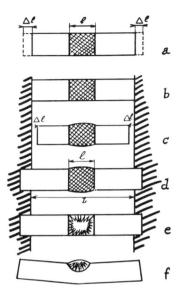

2.23 Creation of residual stresses by local fusion or heating: a) Heating; free expansion and shrinkage $2\Delta 1 = 1\alpha \Delta \theta$; b)c) Heating; hindered expansion, free shrinkage; d) Heating; hindered expansion, restrained shrinkage $-2\Delta 1 = \frac{R_e}{E} \times L$ (R_e yield strength, E module of elasticity); e) Localised fusion over the whole right hand section, restrained shrinkage; f) Dissymmetrical fusion; auto-restrained shrinkage.

c) During cooling, the heating zone shrinks, as in a), but the final length is reduced by $2\Delta 1$, which has been absorbed by the plastification of the heated zone. When this experiment is carried out, we see that the bar falls down because it has become too short to be held by the walls, but it is not the seat of any residual stress because if expansion has been prevented, shrinkage has not.

d) If we repeat the experiment but with a bar which is this time secured by its ends to the two walls, i.e. incapable of any overall movement due to expansion and shrinkage, heating results in the same plastification phase as before; but, on cooling, the shrinkage which led to shortening of the bar is prevented. Consequently, the bar is subject to stress which peaks on completion of cooling when the elastic limit has returned to its initial value.

If we free the bar after cooling, the stress, which was at the most equal to the elastic limit, is spontaneously released by elastic shrinkage and the bar becomes shorter. If the part remains clamped, only a later treatment will relieve the stress.

e) We can repeat experiment d) above, but with continuation of heating until we obtain a fusion zone occupying the full bar section. In this case, an additional factor is involved at the moment of cooling, namely the shrinkage which accompanies solidification. Contrary to what is often stated, this shrinkage does not result in stress, because at the stage when it intervenes, the metal which is solidifying or has just solidified offers practically no resistance to the dimensional variation caused by the solidification shrinkage. After full solidification, the bar lends itself to shrinkage, first by its own plastic deformation,[17] then by elastic deformation, having reached a lower temperature, at which its elastic limit becomes appreciable. In this case, the maximum value of residual stress is necessarily equal to the lowest of the two elastic limits, that of the parent metal or that of the weld metal.

f) The cases examined up to now have concerned even heating or fusion throughout the whole section, so that there was no angular deformation of the bars which, even though free, remained rectilinear. In reality we get a result very close to that for single run welds affecting the entire thickness such as electroslag welds or electron beam welds, or, to a certain extent, electric arc multi-run welds on narrow gaps. As soon as there is dissymmetry in the arrangement of affected or weld zones (weld bead illustrated in case f, or angular groove), we see the appearance of an angular deformation of the part assembly (if it is not prevented by effective clamping) which is the seat of a field of residual stress. Subsequent heat treatment will relieve the stress but not eliminate the deformation. This demonstrates that, even for a completely free part, the fact that the heated or weld zone does not affect the whole of the straight section leads to hindrance of deformation on heating as on cooling, hence residual stress and deformation.

To sum up, we can list the essential characteristics of the formation of residual stresses in welding as follows:
- residual stresses result from hindrance of heat shrinkage on cooling, following plastification associated with localised heating;
- residual stress appears from the beginning of cooling and reaches a maximum when cooling is complete (Fig. 2.24);[18]
- the more deformation is prevented, the higher the value of residual stresses, i.e. the greater the clamping, as defined below. They do not exceed the elastic limit of the metal they affect.

The important role played by residual stresses, in particular vis-a-vis the risk of cracking, and the dependence of the stresses themselves regarding constructional arrangements of the parts to be welded, have led to the quantification of what has been called restraint, i.e. the extent to which the hindrance of expansion and shrinkage is likely to generate residual stress. Such a quantification makes it possible at the same time to evaluate the representativeness of crack tests compared with practice. This is the reason why Japanese researchers proposed the definition of the restraint intensity for a butt weld as being the transverse force which must be exerted on the edges to be welded to prevent distortion and maintain the original geometry. Thus defined (and expressed in N/mm/mm), restraint intensity, evidently, depends on the modulus of elasticity of the material, but also on the thickness and free length l[19] (Fig. 2.25) liable to become deformed elastically under the

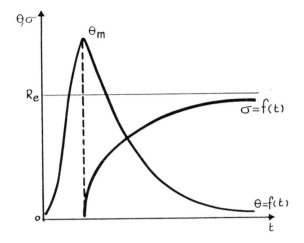

2.24 Establishment of residual stresses, from maximum temperature (nil stress) to cooling (maximum stress, at most equal to yield strength R_e).

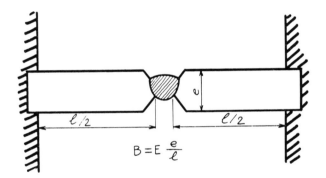

2.25 Definition of restraint intensity.[*]

effects of shrinkage on either side of the weld bead. This quantification has made it possible to evaluate various types of assembly from the point of view of restraint and, subsequently, to adjust for each one the necessary severity of the corresponding weldability tests. Thus, to give an example, for welding the web of the girder illustrated in Fig. 2.26, the restraint intensity is 6500 N/mm/mm if l = 150 mm and 4000 N/mm/mm if l = 300 mm.

---

[*] Doc.IIS/IIW-536-77: Japanese work on restraint intensity in relation to cracking of welds. Soudage et Techniques Connexes, no. 1/2, 1978, pp 61–75.

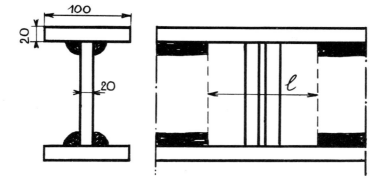

2.26 Practical example of restraint: addition of a girder web.

Possible effects of residual stress and deformation

The residual stresses and deformations generated by welding are liable to participate in the following phenomena and performances:

During welding or shortly after:
– hot cracking;
– cold cracking;
– ageing;
– lamellar tearing.

During fabrication, after welding:
– cracking during reheating;
– dimensional instability during machining.

In service:
– mechanical performance (buckling);
– participation in breaking process (brittle or through fatigue);
– stress corrosion;
– dimensional instability.

The essential role played by deformations, stresses and lamellar tearing will be considered in this chapter. Other behaviour will be studied later, in the context of the metallurgical phenomena which also condition them.

Stress relieving

Given their possible effect, it is often advisable and sometimes absolutely vital to eliminate residual stresses by means of a treatment called relief. Whether this treatment is thermal or mechanical, it acts by the same mechanism which consists of permitting the plastic flow necessary to relieve the parts under stress.

Thermal relief uses the fact that the yield strength of a material, in particular steel, diminishes when temperature increases, so that the zones under stress in the elastic condition at ambient temperature plasticise gradually as temperature increases, thus giving rise to a more or less complete relief depending on the temperature reached. More details are given in Chapter 9.

Mechanical relief implies cold deformation. This can be obtained by putting the welded part under stress at a rate liable to cause plastification, at least where stress 'peaks' exist.[20] In this way we can treat pressure vessels in the as-welded state by putting them under hydraulic pressure. In this respect, French pressure vessel regulations allow a pressure vessel to be exempt from heat treatment if it has been tested to 85% of the yield strength.

Another, less widespread, practice is that of *vibration relief*: the welded parts are subjected to vibratory stress, either as a whole, or locally, by means of pulsators, so that a resonance state is established in the zones containing tension points and brings about the necessary reduction of residual stresses.

Finally, we shall mention the *'Linde' process*, which consists of step by step flame heating of two strips situated on either side of the weld to be relieved which, remaining cold, undergoes plastification associated with the expansion of the heated zones.

Lamellar tearing

Definition

The information above on the phenomena of expansion and shrinkage in welding and their effects in terms of deformation and stress enable us now to deal with a troublesome consequence, liable to occur during steel welding, namely lamellar tearing, which, as regards welding, is essentially associated with its thermomechanical effect.

We use the term lamellar tearing to describe a type of crack stretching parallel to the fusion boundary of weld beads (essentially fillet welds), and, in the main, parallel to rolling surfaces. Inspection after breaking, either accidental or deliberate, shows that the fracture presents a stepped band with glossy tags parallel to the rolling plane, joined by breakage zones with deformation, roughly perpendicular, joining these tags (Fig. 2.27). A more detailed inspection shows that this aspect is caused by decoherence at the interface between inclusions flattened by rolling (essentially sulphurous) and the material containing them. These decoherence zones, which are the reason for the lamellar appearance of the break ('onion skin') are joined by steps corresponding with the breakage of the matrix.

This type of cracking, which results from a weakness in the material when it is stressed in the 'short traverse' direction, i.e. perpendicular to the skin of the rolled product, can occur at welded joints whose geometry encourages the predominance of stresses in this direction. The cracks can occur in the heat affected zone or in the

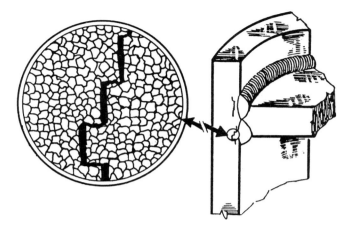

2.27 Practical instance of lamellar tearing, at the fillet weld of a diaphragm inside a ferrule: macro and microscopic appearance.

unaffected parent metal. They do not seem to result essentially from a metallurgical process.[21] If they are not detected at the time of construction (ultrasonic inspection is the most advisable), they can develop in service or even be the source of other failure processes, fatigue in particular.

The lamellar tearing mechanism is revealed by the Cranfield test (Fig. 2.28), in which the shrinkage of a specially prepared multipass fillet weld provokes a stress (and movement), by a lever effect exerted on the skin of the bottom element, hence the tearing effect generated by the movement of the top element towards closure of

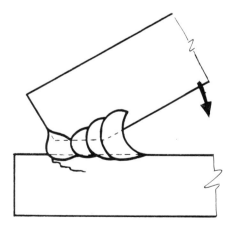

2.28 Principle of the Cranfield test.

the angle with the bottom element. If this movement is halted, for example by means of gussets holding the two elements securely before welding, there is no tearing. This shows that, contrary to what is often said, this phenomenon is not associated with restraint and brittleness (as is the case for cold cracking) but is connected with the deformation ability of the material in the short traverse direction.

Factors in lamellar tearing

The aforementioned explanations show that the risk of lamellar tearing is associated on the one hand with the quality of the parent metal, and on the other hand with the geometry of the joints (design) and with the welding sequence; but its occurrence is also dependent on the metallurgical characteristics of the weld metal.

With regard to the quality of the parent metal, the discovery of the role played by flat inclusions[22] of rolled products has led to the rarification of those inclusions[23] by the increased purity of steel, especially as regards sulphur content. Also, the correlation between the risk of lamellar tearing and mechanical performance in terms of the short traverse has caused the contraction measured on a traction bar (possibly composite) to be deemed a criterion when taken in that direction. Hence the appearance in iron and steel standards of Z steels which provide a guarantee based both on purity and short traverse contraction. Unfortunately, the use of Z steels means a relatively high additional cost; so their use is reserved for constructions in which the risk of lamellar tearing or potential danger of its consequences are particularly high. In other cases, it is preferred to act via other factors of the phenomenon and, when necessary, select products by looking for flat inclusions by ultrasonic inspection (in particular by focused beams), especially in the area where sensitive welds are to be located.

As for the design, which is implicated because lamellar tearing is caused by welding induced stresses acting perpendicularly on product skins, it is desirable to avoid arrangements which lend themselves to that effect (Fig. 2.29), and also preparation imperfections (abnormal spacings) which reinforce the problem. Where there is doubt, suspect regions can be 'buttered' before the actual welding is carried out (Fig. 2.30). The design being set up, the sequence of constituent joints of a welded assembly also plays a role, the general principle being to balance as far as possible the stresses due to welding of a given joint.

The filler product is involved in the mechanism of lamellar tearing via the yield strength of the deposited metal. Indeed, as we shall see later (Chapter 4), the necessity of deoxidising the weld metal leads to the incorporation of appreciable quantities of deoxidating elements (in particular Mn and Si) which also have the effect of raising the tensile characteristics of weld metal in one or more runs, in particular the yield strength. As already mentioned, the stresses established during cooling of a weld are of the order of the yield strength of the material being welded. If therefore that yield strength is greater than the tensile strength of the base steel in the short traverse, tearing is unavoidable. So it is desirable to select the welding

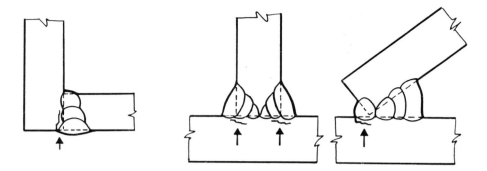

2.29 Preparations to be avoided vis-a-vis lamellar tearing.

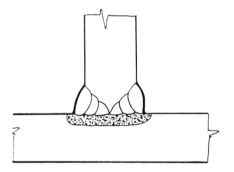

2.30 Buttering to avoid lamellar tearing (see description of buttering in Chapter 6).

process and the filler products to give the lowest possible yield strength in the weld metal. There is even a case for doing welds in which the first runs are as soft as possible with subsequent runs being harder.

Some confusion has been introduced into the explanation of lamellar tearing by the discovery that this phenomenon is reduced, if not prevented, by preheating. This is true, but not in relation to hardening or diffusion of hydrogen. More simply, preheating increases the temperature between runs and duration of cooling; so it enables us to exploit the plasticity of the weld metal in a temperature range where its yield strength is not as high as it is at ambient temperature. If necessary, it is possible to make use of an intermediate stress relieving heat treatment during welding.

Chapter 2 **Comments**

1 This does not mean to say that heat distribution in the welded metal is independent of the energy distribution at the heat source. We shall see it, for example, in comparing electric arc and electron beam.
2 We shall observe that here the word 'simulation' is used in the literature to designate two

different procedures: the first, which is the oldest, consists of imitating the welding heat cycle from experimental data by means of a suitable heat source (Joule effect, induction, light radiation); the second tends to reconstitute the welding heat cycle by calculation using computer program, on the basis of operative conditions.

3 There are also commercially available pencils permitting industrial utilisation of this principle, for example for control of preheating temperatures.

4 In anticipation of what is to follow, we can note that, given equal thickness, this distance is a dependent variable of energy input.

5 This is only true for a part of undefined dimensions. A part with smaller dimensions reaches a final uniform temperature which is higher than initial temperature.

6 Strictly speaking, thickness must be involved. We could accept, for what follows, that the heat solid presented concerns a weld bead on thin metal plate, or even heat distribution in a plane parallel to the surface of a thick metal plate.

7 For example in resistance spot welding, butt welding with different heat sources or even arc stud welding.

8 French standard A36-200 suggests that evaluation of nominal energy should be based on the length of electrode consumed to apply a given length of weld bead.

9 This modification of arc characteristics results in an accentuation of the volatilisation phenomenon in weld metal, which poses an additional problem of professional hygiene.

10 From which results the possibility, in a case where moderate preheating is necessary, of being satisfied with an initial preheating prior to the first run, without the necessity of maintaining it later, that temperature rise taking over from the preheating. The decision to be made depends on the run sequence.

11 The gas welding heat solid is rounded, whilst that for other processes is generally pointed, except in the case of electroslag welding.

12 As in EB welding, the cooling speed slows down when thickness increases.

13 Which obviously, given the very large thickness which can be welded in a single run, does not mean to say that it does not permit fast welding. Nevertheless, cooling speeds are expressed in tenths of degrees per minute.

14 With vertical beam welding on a horizontal plate, this liquid cylinder is maintained by the surface tension of the weld metal, which defies gravity. If the weight of metal becomes excessive, it collapses and welding becomes impossible. Hence the limitation on weldable thickness in this position.

15 We shall see later that, among these phenomena, the expansion associated with martensite forming is itself a stress generator.

16 These experiments take account only of the transverse stresses due to welding. In reality, the same phenomenon is also involved via a longitudinal component.

17 If the elongation capacity of the weld metal is not sufficient to permit the deformation imposed on it, there will be hot cracking.

18 This explains the delayed qualification attributed to cold cracking, which we shall see later, effectively occurs after welding. As to postheating, we shall see later how it acts by suspending the creation of stress (Chapter 7).

19 A simple calculation shows that restraint intensity is proportional to thickness and inversely proportional to restrained length.

20 Localised treatments like the use of rollers, hammering, shot peening, etc, recommended for improving fatigue performance, are not relief treatments. On the contrary, they produce the improvement by creating superficial localised compression stresses in the zone to be protected.

21 A certain possible effect of hydrogen has been mentioned in the literature, but not in a decisive manner, except in the case where lamellar tearing is associated with cold cracking.

22 Inclusions are more harmful the thinner they are. Hence the influence of the reduction of thickness when rolling and, in this respect, the advisability of continuous casting compared with casting into ingot moulds.

23 Another approach consists of keeping the inclusions in the globular state and rendering them less deformable by means of special additions, for example cerium.

3 Introduction to the metallographical examination of welds

Characteristics of metallographical weld inspection

General points

Metallographical inspection, in association with the investigation techniques which complete the information supplied by the metallographical examination, is the basis of the metallurgical study of welds. So it is necessary, for macrography and micrography, to examine the characteristics of these two techniques when applied to weld examination, for the purposes of research, development, quality control and also education.

Except when it is performed non-destructively on replicas (see later), rarely for macrography, more often for micrography, metallographic examination is carried out on samples. It is therefore destructive or at best semi-destructive. Whenever possible, it is preferable that these samples should include all zones of interest as regards the weld under examination, the only serious obstacle being the weight and bulk of samples (thick products), especially in micrography. In welds carried out with a mobile heat source, the samples most often inspected are taken perpendicular to the weld line (Fig. 3.1a). As we have seen previously, such a sample is representative of the whole weld portion where the quasi-stationary conditions have been achieved during the welding operation. Taking samples from other positions or directions is not however excluded; in particular, we obtain interesting information from longitudinal sections (Fig. 3.1b) especially with regard to the morphology of any cracks on which judgement can be made only after examination of the two types of sections, transverse and longitudinal.

In the case of welds made without movement of the heat source, specimens are generally taken in accordance with a symmetry plane; such is the case, for example, for a resistance spot weld, examined in a meridional plane (Fig. 3.2a) but, here again, it may be necessary to examine another section, for example an equatorial plane (Fig. 3.2b.)

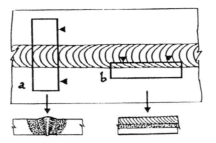

3.1 Orientation of sampling and polishing plane for macrographs of welds made with movement of heat source: a) Transverse section; b) Longitudinal section.

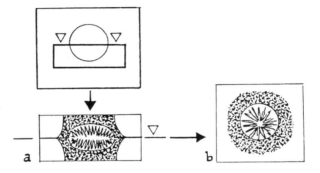

3.2 Orientation of sampling and polishing plane for macrographs of resistance spot welds: a) Central section; b) Equatorial section.

Sampling and preparation techniques

Except when carried out in connection with expert appraisal operations where it is done on samples, macrographic examination takes place on specially produced weld specimens, or even on test pieces used elsewhere for other tests (hardness tests, mechanical or even weldability tests). Macrographic examination has also in the past been used as a means of quality control – with the reservations that we shall find later – but with the obligation of having to mutilate the construction in question to extract samples, so that the development of more suitable means of inspection led to this practice being abandoned. Nonetheless, it remains a method of sampling, still practised today by means of a circular cutting tool[1] which takes up boat shaped specimens (Fig. 3.3); the profile of the zone thus mutilated lends itself better to the vital subsequent repair by welding than in the case of drilled cylindrical samples. This type of sampling is used as a complement to non-destructive testing (radiography or ultrasonic examination) when it is necessary to specify the nature and position of defects detected and, where necessary, to evaluate the possibilities of repairing them.

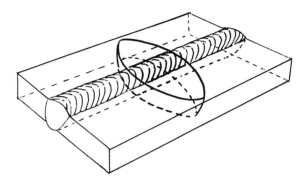

3.3 Boat-shaped semi-destructive spherical drilled trepanning of a sample.

Whatever the object of the examination, it is vital that the sampling and preparation techniques used do not cause heating liable to modify the macrographic or micrographic appearance of the zones to be inspected. If thermal cutting has been used to obtain a sampling blank, the dimension of that blank should be sufficient to permit the area spoilt by the cutting heat to be largely removed by mechanical means.

The same precaution is valid for straightening and polishing of samples, at least mechanical polishing, whether for macro or micrography; the difference between the two lies only in the degree of polishing intended in view of the necessary etching to show the constitutional and structural variations sought.[2] Electrolytic polishing, carried out essentially for micrography, makes it possible to attain a very high degree of polish without any impairment or cold deformation of the surface to be inspected.

Possibilities offered by macrographic examination

Macrographic presentation of a weld – definitions

The chemical etching of a sample for macrography reveals, amongst other things, the constituent zones of the weld, which appear via the intensity of etching or different colorations. These differences reveal the variations in constitution and structure generated by the welding operation. If the etching is strong enough, these differences will be visible to the naked eye or under low magnification. For example, in the general case of a butt weld with fusion achieved in a single run, we see the appearance of the following zones, marked on Fig. 3.4a.

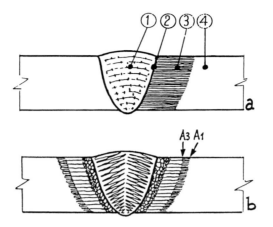

3.4 Constituent zones of a single pass fusion butt weld: a) General case, 1 Weld metal zone, 2 Fusion boundary, 3 Heat affected zone, 4 Parent metal; b) Particular case of steel.

Weld zone

This is the liquid region during execution of the weld. It is obtained by fusion of the parent metal, with a significant contribution of filler metal. If the weld is as welded, the grains resulting from solidification are generally visible without magnification. The metal which constitutes the weld is called weld metal. The metal added to the liquid state, i.e. before dilution, is called deposited metal.

Fusion boundary

This zone corresponds to the limit up to which the parent metal has been induced to melt. It also marks the border of the fusion zone and is evidenced by the difference in structure which we can see on either side. If we look more closely, we must take account of the fact that any alloy (in the widest meaning of the term, i.e. metal/metal or metal/impurity) is characterised by a temperature range called solidification range in which both the liquid and solid phases coexist. Also, another point of interest of this zone is that it is the seat of the initiation of the solidification of the molten metal. We shall return later to the consequences of these two characteristics (see Chapters 4 and 5).

Heat affected zone (HAZ)

As we have already seen, the HAZ is the zone in which the welding heat cycle has produced one or more transformations from the initial state to the solid state involved in heating.[3] As each transformation is characterised by a minimum temperature, the outer limit of the corresponding affected zone coincides with the isotherm which is characteristic of that temperature; that isotherm more or less clearly depending on whether or not the phenomenon is progressive in relation to the temperature. Thus, on Fig. 3.4b relative to steel, we can by a progressive colour variation, distinguish the entry into the area where the temperature exceeds point A_1 (progressive dissolution of the pearlite); we can distinguish the entry into the area fully austenitised by heating over the A_3 point by means of a clear discontinuity. Nearer to the weld zone, the enlargement of the austenite grain, a progressive phenomenon, has generated an area of overheating more obvious, but not marked by a clear isotherm.

Parent metal

Although obvious, it is important to mention the parent metal as being part of a weld sample for macro or micrographic examination. First of all, the parent metal serves as a reference for evaluation of the transformations occurring during welding. Also, certain modifications are not revealed by macrography because they involve constituents and structures too fine or too dispersed for the corresponding heterogeneity to be revealed by the macrographic process. These structures are on the optical microscopic or even electronic scale. It should not therefore necessarily be concluded that the parent metal is not affected because the macrographic examination has revealed nothing. This is the case, for example, with the steel ageing caused by a welding operation close to a cold deformed zone.

Specific cases

For welding processes in which melting does not occur and those where a liquid phase is eliminated, there is obviously no weld zone and we see only one fusion boundary on either side of which there is only metal which has remained in the solid state during the welding operation, but which may retain the pattern of the deformation it has undergone, hot or cold, depending on the process studied. So it is, for example, for flash or friction welding of steel (Fig. 3.5), the macrographic procedure reveals simultaneously the heat affected zone and the deformation suffered by the metal, marked by the deviation of its fibres. We could also mention macrographic peculiarities relative to other processes such as brazing or braze welding, diffusion welding, etc.

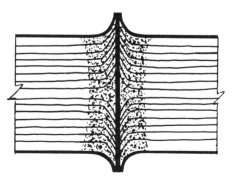

3.5 Macrographic appearance of a hot pressure weld (steel): deformation of fibres at fusion boundary.

Interpretation of macrographs of steel welds*

If we limit ourselves to fusion welds made in one or more runs, we can list and comment on the information supplied by macrographic examination, as above, on the one hand on the geometrical or physical anomalies of welds, and on the other hand on their procedure conditions.

As regards observable geometric or physical anomalies, it is clear that only those emerging on the sections examined are detectable. In that case, they can also be observed without etching on sections which can be guided by the results of the non-destructive testing, which are thus detailed and illustrated.

In the context of geometric defects we can mention:

- linear misalignment;
- angular deformation (roof effect);
- weld beads offset on face and reverse side;
- undercuts;
- excessive reinforcements, or lack of metal;
- lack of root fusion defects.

We can mention the following as examples of physical defects:

- porosity;
- non-metallic inclusions (slag);
- lack of fusion;
- cracks.

Given that the only part of the anomaly visible is the part crossed by the section under examination, inspection of a single section is not enough to decide on the form of the defect observed. This comment is essentially valid for inclusions, porosity and especially cracks which can only be accurately evaluated by examination of a number of sections.

Welding conditions can be detailed by macrographic examination, at least as regards the following elements, valid for steel fusion welds:

- determination of welding process: The question only arises in the case of an expert appraisal, but it is a good thing to remember that the welding process is identifiable, primarily by the profile of the weld zone (if there is one) on transverse sections and by the greater or lesser extent of the heat affected zone.
- number, arrangement and order of runs: Macrography supplies information on this data by observation of the following characteristics, illustrated in Fig. 3.6 for a three run weld, one of which is on the reverse side.
- in the heart of the weld metal, fusion boundaries are visible with their concavity

* H GRANJON: Some information for reading macrographs of welded assemblies, Soudage et Techniques Connexes No. 314, 1950, pp 81–84

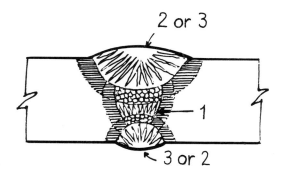

3.6 Macrographic interpretation of the order and arrangement of passes (for a three-pass weld in steel).

towards the side where the runs have been deposited. It is possible to distinguish both order and number from the overlaps of the runs.

– the zones of weld metal of a run, thermally affected by other runs, are visible by their concavity and different structure (refined), which confirms the prior diagnosis.

– the limits of the heat affected zones of the parent metal, resulting from each run, confirm the observations made on the weld metal by their orientation and intersections, at least at the periphery of the weld zone.

– the weld metal of as-deposited run (marked 2 or 3 and 3 or 2) are distinguishable by its structure not transformed by the effect of subsequent runs.

– uncertainties can however remain on the order of runs. So, on the macrograph in Fig. 3.6, we cannot tell whether the sealing run on the reverse side was made before or after the final run on the face side.

Evaluation of dilution rate: According to the definition already given of dilution rate, it is sufficient, for the evaluation of this rate in the case of a deposited bead (Fig. 3.7a) or a build up, to measure[4] surfaces s (molten parent metal) and S (total molten metal) to calculate this rate, equal to $d\% = \dfrac{s}{s+S}$ hence evaluation of the chemical composition of the weld metal from those of the parent metal and undiluted deposited metal. This measurement, very useful in the case of heterogeneous build up (for example stainless steel on low alloy steel), assumes that the procedure and penetration conditions are constant throughout deposition.

The same measurement is possible in a weld, provided that the initial profile and gap traced on the macrograph (Fig. 3.7b) are known. For steel welds, macrography makes it possible to determine or verify whether heat treatment, before, during or after welding has taken place.

Welding on a part in the initial annealed state produces a heat affected zone, clearly marked by isotherms A_1 and A_3 (Fig. 3.8a). If the steel has been treated prior to welding, by hardening and tempering, the heat cycle exerts its effect beyond these isotherms, causing an 'over-tempering' in the zone which has been heated beyond

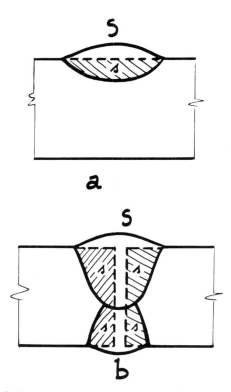

3.7 Planimetric evaluation of dilution on macrographic image provided initial preparation is known: a) On surface deposit; b) On welded joint. S total weld metal surface, s surface of parent metal affected by the weld.

the temperature of the initial temper. The macrographic etching marks this zone with a less contrasted halo surrounding the austenitised zone (Fig. 3.8b). The preheating causes a reduction in the temperature gradient, and a spreading of the heat affected zone. Compared with that of a non-preheated bead (Fig. 3.8c), the macrograph of a bead preheated and deposited with the same energy (Fig. 3.8d) demonstrates this spread, at the same time as a reduction in contrast.[5]

Annealing is carried out at a temperature greater than that of point A_3, to obtain full austenitisation. If such annealing has been performed on the whole of a welded part, or on a sufficient width around the bead, it has caused the disappearance of the HAZ, which has been reaustenitised at the same time as the parent metal, and produces refining of the grain of the fusion zone (which is one of the objectives of annealing) These two consequences make the macrographs[6] of an as-welded bead (Fig. 3.8e) different from that of an annealed bead (Fig. 3.8f). This observation also holds true for hardening and tempering treatments of welded assemblies.

When treatment is carried out at a temperature where austenitisation is not required, as is the case for stress relieving treatments, the affected zone is neither

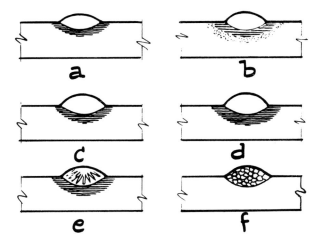

3.8 Identification of any heat treatment prior to, during and after welding, by their effect on the weld metal zone or heat affected zone:
    initial condition: a annealed, b hardened and tempered
    preheating: c without, f with.
    annealed after welding: e without, f with.

suppressed nor modified dimensionally, but the temper effect caused by the operation is apparent by a lesser colour contrast compared with that in the case of Fig. 3.8.

Comments on the miscellaneous uses of macrographic examination

Given the information it supplies, macrographic examination of welds is used for various purposes, and a few comments can be made depending on these uses.

Firstly, macrographic examination precedes and guides the micrographical examination, regardless of the objective of this latter (research, appraisal, quality control or education); even if subsequent operations complete that of the macrographic sample, the use made of that sample is indispensable for localising micrographical examinations and, thereby making use of the conclusions. Also, the results of a micrographic exploration of a weld are often presented at the same time as a macrograph where the positions of the points examined are situated.

Similarly, macrography is indispensable for locating test samples, or at least the blanks. So it is that underbead hardness tests or on actual welds, and the various mechanical tests, including notch impact tests, for which the macrographic process on the blanks themselves is indispensable to specify the position of the notch. So it is also for chemical micro-analyses, in particular by means of the Castaing micro-probe. The results of these various tests are often presented in the form of a graph superimposed on the macrograph of the weld examined.

As a means of inspection, apart from the reservations expressed above about the damaging effect of taking the necessary samples, it should be remembered that macrographic examination takes account only of characteristics emerging on the sections inspected. The result of this is that random anomalies such as the majority of the defects listed above cannot be detected with certainty. These defects can only be reached on localised sections by means of non-destructive testing. On the other hand, procedure characteristics which are established all along the length of a weld can only be revealed by macrographic examination, on any section (by virtue of the quasi-stationary condition). Hence the importance of macrographic examination for developing and approving welding procedures or for verification that those procedures have been properly applied. In the first case, examination is made on representative assemblies; in the second, sections taken from test bars or test plates prepared during manufacture at the same time as the actual welds. In this case, the greatest care must be taken that run-off test plates are truly representative from the point of view of the heat cycle they undergo (welding and treatment).

Complements to micrographic examination

Thus guided by the macrographic examination, the micrographic investigation (optical or electronic) does not pose any specific problem for the welds as regards examination techniques. A few practical indications only can be given.

With regard to the finishing of samples, carried out in the same manner for optical or electronic micrographs by sweeping, the polishing and electrolytic processes present a few difficulties if it is desired to obtain an even polish and effect over all the constituent zones of the weld, which have unequal dissolution speeds. So most often we make do with a few local polishings (each over a few millimetres diameter) in the places we want to inspect, previously located by macrographic etching.

For electronic microscopic examination by transmission, we use replicas, generally in carbon, which are prepared from plastic replicas obtained by polished etched samples.[7] We can also use very thin samples (thinned by electrolyte dissolution); preparation of these is more delicate. In this respect, from the point of view of continuity of observations on the polished surface, micrography by sweeping lends itself better to weld examination than micrography by transmission.

Micrographic examination can sometimes be quantitative, given the necessary additioned equipment. Let us mention, for example, evaluation of delta ferrite in the weld metal of austenitic welds, of the proportion of martensite in the heat affected zone of steel welds, or of inclusions in the weld zone.

Finally, for research or educational purposes, micrographic examination may be associated with cinematography, as has been the case at the Institut de Soudure, for studying the role of hydrogen in cold cracking and the release of this gas from the weld metal.

Conclusions

Metallographic examination of welds constitutes one of the most important techniques in the metallurgical study of welds, for its results make it possible to gather and interpret more fragmentary results given by other investigative techniques. Metallographic examination, as we have just seen, also enables us to locate fundamental phenomena which are going to be tackled later. To that end, we shall develop the three metallurgical aspects previously mentioned (preparation, fusion, heat treatment), in the following order:

– in liquid phase (Chapter 4)
– on solidification (Chapter 5)
– in solid phase[8] during heating and cooling (Chapters 6 and 7)

Each of these headings will be dealt with in a general way, with the help of examples concerning various metals and alloys chosen to throw light upon the more practical information provided by the literature.

Chapter 3 **Comments**

1 The corresponding apparatus is known under the name 'weld prober'.
2 With Portevin, we call that which concerns the nature and proportion of the phases present in the metal or alloy examined 'constitution' and that which concerns the morphology of phases 'structure'. We shall go into this distinction in more detail later.
3 Transformations on cooling are more difficult to locate by macrography because the necessary temperature condition achieved generally adds a condition relative to the law of cooling.
4 This measurement can be made by planimetry or, very simply, by cutting out and weighing on a photographic print.
5 Indeed, the reduction in cooling speed, which gives rise to a change in the mode of transformation results in a less marked contrast in the colouring of the heat affected zone by the reagent used.
6 The disappearance of the heat affected zone under the effects of the annealing as a criterion of the effectiveness of the operation does not always require a sample to be taken: for example, for liquid gas bottles, inspection can be carried out by etching the surface of weld beads after a brief preparation (French standard M88-703).
7 The same technique is applied for microfractography, i.e. examination of cracks or breaks. The replica is then taken directly from the fractures being examined, usually without etching.
8 In fact, it is better to give up the chronological order and make an unbroken study of the solid state transformations, whether they concern the parent metal or the weld metal.

4

Formation of the weld metal zone

Summary of the causes of evolution of the chemical composition of the weld metal

From the first chapter, during our study of the general characteristics of the welding operation, we have mentioned all the phenomena responsible for developing the chemical composition of the weld metal during welding using a process in which melting is involved. So we can effectively talk about welding as a metallurgical operation, but on condition that account is taken of the particular conditions under which that operation is performed.

To study the chemical evolution of the weld metal zone, we shall examine the following:
– the physico-chemical behaviour of the elements present in the liquid phase (filler or parent metal), which can be subject to volatilisation or may react against each other.
– the influence of the surrounding medium, whether gaseous (atmosphere or other gases), liquid (slag), or solid (parent metal).
– the particular case of the fusion boundary.

Physico-chemical behaviour of the elements present

Volatilisation

This is liable to occur during welding, for a given element, each time that element, alone or in an alloy, is raised, in the liquid phase, to a temperature at which its vapour pressure reaches a significant value. The classic example is zinc, an element which melts at 419 °C and can be gas welded without problem,[1] because the molten metal is in thermal balance with the parent metal, i.e. at a temperature where vapour pressure is low. On the other hand, alloyed with copper to constitute brass which melts at approximately 920 °C, zinc volatilises abundantly at that temperature, hence the appearance of zinc oxide fumes[2] (by oxidation of the zinc vapour on contact with the air) during gas welding, or also during TIG welding, even more so because the temperature is higher on impact of the arc.

It is because of the increasing temperatures they induce that welding processes are classified from the point of view of tendency to volatilise, but account must also be taken, of soaking and of surrounding pressure. Indeed, the increase of this pressure causes constriction of the electric arc, hence an increase in temperature which generates a more intense volatilisation. This poses a problem of respiratory hygiene with hyperbaric steel welding because of the more intense release of iron vapour than at atmospheric pressure, where it is already susceptible. Paradoxically, the reduction of pressure also means increased volatilisation, this time because of a direct effect on the phenomenon. This comment is applicable to EB welding: the walls of vacuum chambers are quickly coated with a thin metal layer the nature of which depends on the materials being welded. Normally, the volatilisation phenomenon is limited, because of the rapidity of the welding process, although it is much greater in MIG welding.[3] Nevertheless, it should be taken into account for some alloys such as Al-Zn or Al-Zn-Mg: volatilisation of the zinc modifies the chemical composition of the weld metal zone and may, moreover, affect the function of the gun by spoiling the filament.

Returning to atmospheric pressure welding, we can, in addition to the zinc already mentioned, cite the metals which constitute the base or addition elements of alloy or non-alloy steels, such as manganese, chrome, nickel, aluminium, or even titanium. Of course iron in the Form of Fe_3O_4 constitutes the largest part of arc welding fumes because the vapour given off immediately oxidises on contact with the oxygen in the air. The only preoccupation concerning emission of iron oxide fumes concerns respiratory safety of the operators.

Manganese, a normal element of steel (de-oxidation and desulphurisation) is also subject to volatilisation during arc welding, which means compensating for the corresponding loss through electrode coatings, by means of ferro-manganese. With gas welding in gaseous conditions, the loss is taken into account by a relatively high content of the filler wire, especially in MAG welding where, as we shall see later, an oxidation effect by the protective gas is added to that of the volatilisation. The loss of manganese is much higher for steel with 13% Mn, used as surfacing filler metal for its wear resistance. Here again there is the problem of respiratory hygiene because manganese oxide fumes are harmful to the human body.

Among the steel deoxidising elements, silicon does not pose any problems of this kind, whilst aluminium is susceptible to volatilisation so that deoxidation by aluminium is not usually carried out for filler products. Amongst steel additives, chrome is the most susceptible to volatilisation, as it also is to oxidation, so that the resultant chrome loss has to be made up by wire composition or coatings. Nickel is also susceptible to volatilisation, but to a lesser degree. In addition, the problem of volatilisation arises not only for low alloy steels, but also for alloy steels of the Cr-Ni stainless type. Furthermore, some of these steels contain titanium as a stabiliser element because it fixes the carbon, thus preventing precipitation of chrome carbides liable to corrosion. Titanium is volatile and, during arc welding, the resultant loss in the weld metal destabilises it and renders the weld metal zone prone to corrosion: hence the recommended precaution of using a filler metal which is very low in carbon, or even stabilised with niobium which is not volatile. But this solution is not without disadvantages from other points of view (hot cracking and embrittlement in particular), including that of price.

Other metals are susceptible to volatilisation during welding, either directly as is for example copper which poses problems during weld surfacing using the MIG process, or as a coating.[4] Indeed, welding coated products heats the support material as well as the coating, which may be either broken down by the heating, or volatilised. The oldest case is that of lead, incorporated in the oxide state (red lead oxide) in paints constituting the undercoat of protective coatings: heat breakdown of lead oxide (Pb_3O_4) causes the formation of lead which volatilises and creates a serious danger for the operators (lead poisoning), whether it is welding or thermal cutting. Nowadays, the problem no longer arises for welding because the red lead oxide has been replaced by iron oxide, but it remains where old buildings are demolished by thermal cutting, which, like repairs, requires special operator protection.

As a second case, we come back to zinc, which is involved in iron and steel products, either in the constitution of coatings (e.g. by galvanising) or as a component of protective coatings (zinc paints) applied to products after shot blasting. The following comments can be made on these two situations.

For zinc coated products, fusion welding without precautions causes volatilisation of the zinc, not only in the weld metal zone, with the result that porosities develop on solidification, but also on either side of the joint, where the surface zinc has disappeared over a greater or lesser width depending on the process and procedures used (Fig. 4.1a). It is therefore necessary, subject to the problem of operator protection, to grind the edges before welding to remove the zinc coat which is liable to become incorporated into the weld metal, then to reconstitute the protective coat which has disappeared from the parent metal, by means of an application of zinc paint to the surface and, if necessary, on the reverse side of the fusion zone (Fig. 4.1b).

The zinc paints, used for the preparation of pre-painted products, are made up of an organic binding agent and a certain quantity of zinc powder. The welding heat effect causes the breakdown of the binding agent and volatilisation of the zinc so that, in addition to the precautions taken for the preparation of these paints (nature of binding agents and proportion of zinc), the thickness of the paint coat should be limited to avoid the disadvantages mentioned above. It may be necessary to reconstitute the coat of paint after welding; also, it is advisable to carry out a compatibility test on the filler product, in particular from the point of view of the risk of porosity.

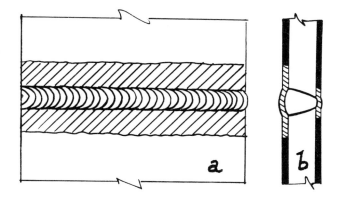

4.1 Fusion welding on zinc coated steel plates: a) Disappearance of the zinc coating by volatilisation on either side of the weld bead; b) Reconstitution of the zinc coating after welding and protection of the weld metal zone.

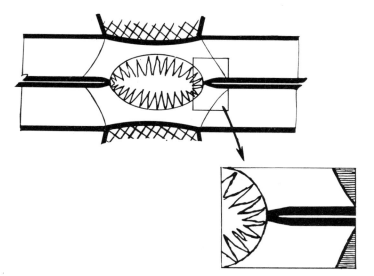

4.2 Behaviour of the zinc coating during resistance spot welding of coated products on contact with the electrode tips and between surfaces to be assembled.

With resistance spot welding, there is no volatilisation; the problem is different because the contact resistance between the zinc coatings of the assembled parts produces rapid melting of those coatings, the liquid zinc being pushed to the periphery of the spot which then forms. The real problem lies in the deterioration of the electrode tips, which calls for special precautions (Fig. 4.2).

Chemical reactions in the liquid state

This section is mainly devoted to the reaction between the iron oxide (FeO) on the one hand, and the steel carbon on the other hand, together with the possible involvement of deoxidising elements, in particular silicon and.or aluminium, when preparing steel. The study of this reaction should start at this stage, because what occurs in the molten steel pool is only a repetition of what might have occurred at the preparation stage in the steel works.

Rimming steels – killed steels – consequences in welding

In the liquid state steel, in iron-carbon alloy, because of the conditions of its preparation, contains a certain amount of FeO which, dissolved in the liquid phase is in chemical equilibrium with the carbon in accordance with the reaction: $C + FeO \rightleftharpoons CO + Fe$.

Cooling modifies the stability of the equilibrium by encouraging this reaction in the direction which supplies the carbon monoxide, which is released and leaves the liquid phase, or indeed finds itself locked in by solidification. This release from the liquid causes agitation of the liquid; that is why the steel which gives rise to this phenomenon is called rimming. Blocking the release of the carbon monoxide by solidification results in porosity, which we see in the ingot, but which, normally, flattens and rewelds during subsequent rolling. Also, because of the agitation of the liquid phase caused by the release of the gas, the impurities, in solution or in the form of inclusions, together with the carbon, cluster together and enrich this phase so that the central section of the ingot, which is the last to solidify, is richer in carbon and impurities than the outside which solidified first and which is less rich in these elements C, S and P. This is the phenomenon of segregation which, later, after hot rolling, results in a chemical heterogeneity of the finished products, richer in carbon and impurities (in particular S and P) in the centre than on the skin (Fig. 4.3a).

If, during production or at the moment of casting, the steel is given a dose of elements which, being oxygen absorbent, avoid oxidation of the carbon (silicon, which oxidises in the form of SiO_2, aluminium, which gives Al_2O_3, or even titanium or zirconium), carbon monoxide is not released and solidification is calm and quiet because there is no bubbling of the liquid; we then say that the steel is killed. Such calm cooling of the liquid phase makes it possible to decant impurities and carbon towards the top (head) of the ingot, which has the effect of producing throughout the biggest part of that ingot a steel which has a uniform distribution of carbon and impurities; this, after rolling, supplies products which are themselves uniform because they originate from sections free from segregation,[6] subject to suitable cropping. This segregation process is called major segregation, as opposed to minor or dendritic segregation which we shall meet again with regard to weld solidification (Fig. 4.3b).

Steel effervescence makes it impossible to use continuous casting, i.e. direct rolling of products without passing through the ingot mould. Hence the quasi-disappearance of rimming steel from products supplied by modern steel works. Nevertheless,

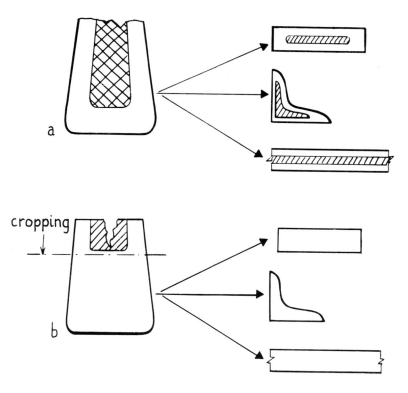

4.3 Influence of the method of steel preparation on solidification in the ingot mould and resultant macrographic appearance for rolled products: a) Rimming steel; b) Killed steel (cropping of ingot removing the shrinkage cavity).

we cannot ignore the characteristics of the welding of rimming steels which the user may meet during modification or repair work, or in countries where the use of continuous casting is not usual.

As a first consequence of the rimmed or killed character of steel, we should mention the fact that, as soon as melting is obtained in welding, the effervescence, which, as we have seen above, has been interrupted by solidification into ingots, appears again. We could say that the welding 'wakes' the effervescence so that, if no deoxidation effect occurs to kill the molten pool, carbon monoxide will not fail to appear and cause the formation of porosities during subsequent cooling and solidification. Thus, in the case of a fusion line without filler metal (thin plates in rimmed mild steel, hence deprived of silicon or aluminium), oxyacetylene welding, characterised by a reducing flame, will supply a sound fusion zone, whilst the TIG process, characterised by an inert atmosphere, will supply a porous bead. The same operation, carried out on killed steel, does not give rise to the same difficulty.

Another consequence of major segregation and hence the chemical dissimilarity of rimmed products lies in the fact that, depending on the location of the welded joint, the welding heat cycle may affect a zone rich in carbon and impurities, originating in the segregated section of the original ingot, or even a pure section with low carburisation coming from the cortical zone or the base of the ingot. We shall see later that there is a big difference from the weldability point of view. So we must handle with care the configurations which, during the welding operation, favour direct action on the segregated zone of a rolled product, for example, in the particularly awkward case[7] of a flange machined from a rimmed steel rolled product for want of a forged product (Fig. 4.4).

Killing the molten pool. Consequences

What happens during steel production and solidification also takes place, on a different scale, in the molten pool. It is therefore necessary to deoxidise the molten pool, not only with a rimming steel, but also even with a killed steel, because, the gaseous atmosphere surrounding the molten metal during its transfer in the electric arc (see later in this book) as well as the molten pool itself can be oxidising; hence the need to incorporate additional deoxidising elements such as silicon and manganese, into the molten metal to ensure the steel is killed. The result of this, at least for non-alloy mild steels, is that in arc welded metal (welded with covered electrodes and especially in a gaseous atmosphere, particularly CO_2) we find contents of these elements higher than the percentages normally encountered for the parent steels.

The deoxidising elements, necessarily excessive in relation to this single function, also have the effect of raising the tensile characteristics of the deposited metal (yield and tensile strength), in such a way that it is usual, if not inevitable to find tensile characteristics in the weld metal higher than those of the parent metal. This comment should be borne in mind in several fields, in particular the interpretation of mechanical tests on assemblies, the significance of defects in terms of fracture mechanics, or even the problems of cracking (cold cracking or lamellar tearing) associated with stresses and deformation.

Influence of gaseous environmental medium

Atmospheric gases: necessity of protection

The beginnings of fusion welding have been marked by gas welding, more precisely oxyacetylene welding. In this process, the gases (carbon monoxide and hydrogen) produced by the primary combustion of the acetylene play not only a protective role vis-a-vis the atmospheric gases (oxygen and nitrogen), but also a deoxidising role.[8]

On the other hand, the need for protection is imperative from the beginning of arc welding because of the weak characteristics obtained for steels (in particular notch toughness) when welding in air without protection; the search for better properties was the motivation, at the same time as other effects, for the development of covered electrodes, which supply double protection by releasing gases and liquid slag.

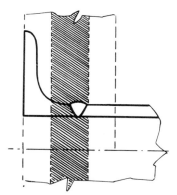

4.4 Assembly of pipe on a machined flange in a rolled rimming steel product: the weld affects the segregated zone.

Later on submerged-arc welding appeared before the use of inert gases made it possible to protect the arc without the involvement of a liquid medium (TIG or MIG processes). Then, with each process taking its place amongst the others, we saw the appearance of welding in an active atmosphere (CO_2 or mixture, MAG process) with or without the involvement of a liquid phase. The last stage is, at present, that of gas free welding, i.e. in the air, suitable formation of the deposited metal being guaranteed by the use of a flux (cored filler wire) or even by the composition of the single filler wire which itself contains all the elements necessary for deoxidation.

These days, the problem of protecting the molten metal also arises for alloy steels, the most usual case being that of Cr-Ni stainless steels. This protection is provided either by a liquid phase resulting from coating the electrodes or flux, or by an inert gas (argon, helium or mixture) during TIG or MIG welding. In the later case, protection must be provided on the back side, by the argon gas or simply by nitrogen.

For non-ferrous metals, the necessity of protection against atmospheric gases is even more vital with the more reactive metals such as aluminium or titanium. For aluminium, the use of a cleaning flux (actually slag-forming) has made gas welding possible (covered electrodes have not been significantly developed) whilst awaiting the appearance of TIG or MIG welding. We now have a mechanism which, in reality, does not on its own provide protection for the molten pool, too oxidisable and especially already oxidised in the metal substrate; it is the electrons which break the aluminium skin which inevitably covers the molten metal, which thus provide the wetting against which this skin is acting. For titanium and its alloys, protection must be stricter because the metal has to be protected as soon as it is heated before it melts, and during cooling after solidification. Hence the use of TIG torches with a 'tail' distributing the protective gas to spread the action, or even 'glove boxes' in which the part is enclosed, or moves through, during welding.

Finally, we must mention the particular case of multiple run welding of materials susceptible to oxidation. Indeed, during TIG or MIG welding, the weld metal of a run may oxidise after solidification when it is no longer gas protected. During the

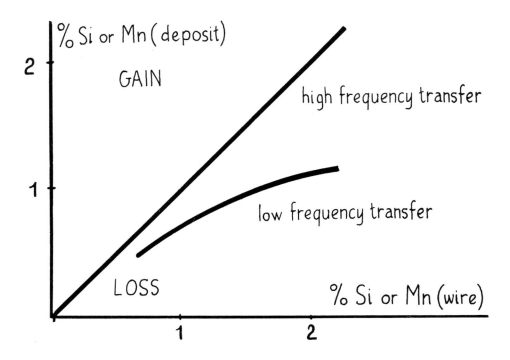

4.5 Evolution of manganese and silicon content in MAG welding (CO_2 protected): role of mode of transfer of weld metal in the electric arc (IIW).

remelt caused during the following run, the superficial oxide thus formed may be incorporated into the molten pool and contaminate it if it is not reduced by an element present in the filler metal. This is, for example, the case with copper, the reducing element being silicon.

Chemical role of the gases active in steel welding

As indicated above, the protective gas used in submerged-arc welding can be either inert or active, at least in steel welding, because the use of carbon dioxide or mixtures of this gas is not recommended for non-ferrous metals and steel alloys. When welding non-alloy steel, the oxidising character of the active atmosphere used causes evolution of the chemical composition (loss of Mn and Si) from the filler metal to deposited metal, an evolution which is itself affected by the transfer mode of the weld metal in the arc (Fig. 4.5).

Influence on the welding operation

The nature of the protective gas also has an influence on the welding operation itself. We have mentioned earlier the role of helium vis-a-vis the heat cycle in the case of caisson welding. This influence also exists for TIG welding in normal atmospheric conditions: helium or an argon-helium mixture, or even an argon-hydrogen mixture permit speedier welding than pure argon, a fact which can be used to advantage for welding alloys sensitive to the heat cycle, for example the maraging type steels.

Also, with gas protected welding, the nature of the protective gas plays a role vis-a-vis the characteristics of the arc and the mode of metal transfer; penetration and consequently dilution are influenced by them.

Standards relating to protective gases and electrode wires

With regard to protective gas, a draft standard A81-010* describes the gases and gaseous mixtures by a series of four symbols relative to:
– the chemical character of the atmosphere created (neutral, ± oxidising, ± reducing);
– the basic constituents of the mixture (argon, helium, carbon dioxide, hydrogen, nitrogen, oxygen);
– the other constituents of the mixture;
– additives not modifying the character of the atmosphere created.

Also, account taken of the influence of the protective atmosphere on the composition and consequently on the properties of the deposited metal, standard A81-311**, concerning filler wire for welding of non-alloy or low alloy steel specifies on the one hand the chemical composition of wire, on the other hand the mechanical characteristics of a deposit carried out in a 15 to 25% carbon monoxide argon atmosphere (in accordance with standard A81-312)***.

Hydrogen

Origin and evolution

It is rare for hydrogen to be intentionally present in welding atmospheres because atomic hydrogen welding is no longer used. Apart from the use already mentioned (mixed with an inert gas), we only find hydrogen being deliberately used in the case of electrodes with total or partial cellulose coatings, containing products so named whose combustion in the electric arc supplies water vapour, itself broken down into its elements: hence the presence of hydrogen, in the form of H+ ions, in the arc atmosphere.[9] But there are other sources of hydrogen

---

* French standard A81-010: Welding and allied techniques – gases and gaseous atmospheres for gas protected submerged welding and for protective atmospheres – Classification, symbols, designation – April 1987

** French standard A81-311: Bare metal electrode wires for gas protected arc welding depositing non-alloy or low alloy metal. Symbols. Filler products

*** French standard A81-312: Bare metal electrode wires for gas protected arc welding. Wire electrodes depositing a non-alloy metal or a metal with low quantities of additional elements. Chemical analysis, mechanical, geometrical and physical tests

which can appear for other types of covered electrodes or other welding processes, always through the breakdown of water in the electric arc. This can be present in the form of crystallisation water from certain mineral products used in the coatings or fluxes, or by the humidity adsorbed by these fluxes and coatings, during manufacture or later.

Finally, we must mention the effect of a damp environment acting directly or by accident, such as when there is a leak of cooling water from the nozzle during TIG or MIG welding.

Hydrogen arising from these various sources, thus ionised, is soluble in liquid iron – and steel – and this solubility diminishes at decreasing temperature, not only during passage from the liquid state to solid state, marked by significant discontinuity, but also during solid phase cooling.

Leaving aside for the moment the details of this variation, illustrated by the curve in Fig. 4.6, called Sieverts curve,[10] we can understand that rapid solidification and cooling maintain an excess of hydrogen in the liquid then solid metal, giving rise to the formation of porosities at the time of solidification if this gas has time to collect, or the maintenance of hydrogen in the form of H+ ions in the solidified metal in the supersaturated solution.

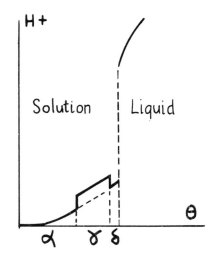

4.6 Hydrogen solubility in iron on the basis of temperature (according to Sieverts): influence of solidification and allotropic transformations.

So the hydrogen, present in the solidified metal at ambient temperature, and diffusing easily through ferrite, tends to escape to the outside through the surface, in relation to time (Fig. 4.7) up to a limit corresponding to the hydrogen called 'diffusible'. This development is accelerated by a rise in temperature: the measurement of

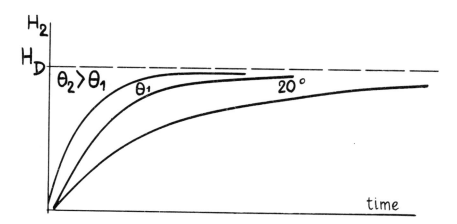

4.7 Pace of hydrogen evolution on the basis of temperature and time from a sample of deposit metal (H_D = diffusible hydrogen).

diffusible hydrogen obtained by heating a weld metal test piece obtained under conventional conditions serves as the basis for the classification of welding products and for the comparison of welding processes and procedures. The result is generally expressed in ml/100g of deposited metal, or weld metal (deposited metal + parent metal), in accordance with convention. To make this clearer, the order of magnitude of hydrogen content observed for the various welding processes and product states is illustrated in Fig. 4.8.

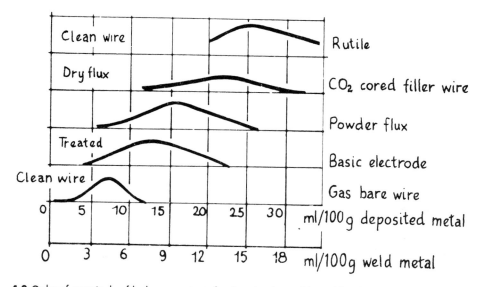

4.8 Order of magnitude of hydrogen contents for deposits obtained by welding (according to Doc. ISS/IIW 343.70).

Returning now to the evolution of hydrogen in terms of time and temperature, as illustrated in Fig. 4.7, we understand that when we wish to carry out mechanical tests on weld metal (mainly tensile tests), it is necessary to establish a reference state obtained by heat treatment of test blanks,[11] the purpose of this treatment being to hasten the release of the diffusible hydrogen, thus avoiding the instantaneous effect (see later) which this gas can have on the behaviour of the weld metal during the tensile test. Without this treatment, the test result could be variable depending on the time separating the welding operation and the test.

Given the importance, among the hydrogen sources in weld metal, of product humidity (which can be measured and translated into 'potential hydrogen'), much importance is attributed to the drying of products when they are manufactured (by heating, often in a vacuum), and to their conditioning, to protect them against becoming damp again inside the packing, especially for basic electrodes, and when they are handled during welding; if necessary, renewed heat treatment should be applied with care being taken where electrodes are used not to spoil the coating.[12]

Finally it must be pointed out that if, for hydrogen, the literature places the emphasis particularly on arc welding with covered electrodes or cored filler wire, the consequences of this gas, in particular from the cold cracking point of view, concern also other processes, for example because of the hygroscopicity of the powders used in submerged-arc welding, or even the humidity adsorption by bare wires in MIG or MAG welding. Indeed, some hardening steels are susceptible to cold cracking at very low hydrogen contents.

Consequences in steel welding

As indicated above, hydrogen in supersaturated solution tends to leave the weld metal and move to the outside. But it can also evolve by diffusion in the solid phase, in the fusion zone as in the heat affected zone. This diffusion, which starts as soon as cooling commences and continues later, is influenced on the one hand by the initial hydrogen content at the end of cooling, on the other hand by the difference in diffusion coefficient between austenite and ferrite, because hydrogen diffuses more easily in ferrite than in austenite (Fig. 4.9). Furthermore, diffusion is encouraged by deformation, elastic or plastic, because the $H+$ ions, of very small size, move by means of dislocation movement.[13] Finally, the consequences of the presence of hydrogen (which we shall have to examine) are associated with the embrittlement of ferrite or martensite, due to the presence of $H+$ ions in the crystalline structure of these two constituents. Ferrite embrittlement is connected to the 'fish eye' formation mechanism during static tensile tests, or 'white spots' during cyclical stresses. Embrittlement of martensite is the basis of cold cracking.

Fish eyes are aspects of fracture that can be seen during tensile tests on machine weld metal test pieces, for example for normalised electrode tests, or welded test pieces, when the break occurs in the fused metal. A fish eye is a portion of brittle fracture on a background of ductile fracture, easily recognisable by its crystalline

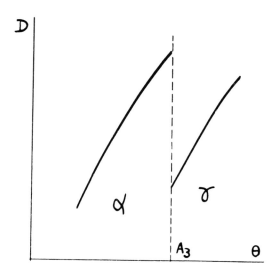

4.9 Hydrogen diffusivity in ferrite and austenite in relation to temperature: D diffusion coefficient.

appearance, surrounding a pre-existing defect such as a porosity or especially an inclusion.[14] This portion of the fracture may be circular or elongated depending on the shape of the defect it surrounds and which is its origin (Fig. 4.10a). The fish eye does not constitute a defect by itself because it does not pre-exist, as is shown by u/s monitoring of a tensile test. We see that the phenomenon only occurs from the maximum point on the effort-elongation curve, i.e. at the moment when the reduction

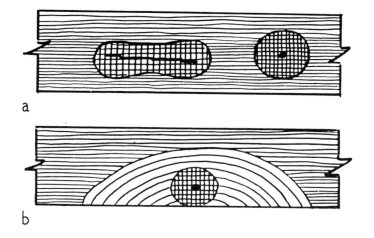

4.10 a) Fish eyes on a static tensile test specimen fracture: brittle ring centred on a defect, ductile fracture; b) White stain on a fatigue fracture: brittle ring, centred on a defect, serving as initiator of a fracture by progressive cracking along a curvilinear front (fatigue) (then plain), involving the entire section.

in area (necking) appears. The result, with regard to the results of the tensile test, is a reduction in elongation and especially in the necking itself. Notch toughness strength may be affected by hydrogen, probably by means of microscopic cracks appearing on cooling after solidification, but not by fish eyes which do not appear during the rapid deformation involved in the impact testing which does not permit the occurrence of the mechanism explained above.[15]

The mechanism of fish eye formation is as follows: by means of the plastic deformation which occurs from maximum effort during the tensile test and increases during the necking phase, the dislocations carry the H+ ions and gather them near the defects where there exists a concentration of stresses. So it is the consequent local embrittlement which produces a brittle fracture around each defect (grains), the rest of the fracture being ductile (nerves) a phenomenon which is accompanied by a necking reduction.

So we can see that a fish eye does not constitute a defect, but that it is the indication of the presence of hydrogen and the result of the combined influence of that hydrogen and defects which already existed. So we understand that we can find fish eyes in the deposited metal from a basic electrode, even with the low hydrogen content which characterises this type of deposit. The low quantity of inclusions, which is another characteristic, results in a more intense entrapment of the hydrogen than if the same quantity of that gas was found in the presence of a larger number of inclusions in a deposit of lesser quality.[16]

The white spots that can be found in the fracture of fatigue test pieces show that cyclical stresses can induce a mechanism similar to that of fish eyes; but, whilst a fish eye needs plastic deformation for its appearance, a white spot occurs, during cyclical stress, for stresses below the elastic limit; so the white spot plays the role of initiator in fatigue fracture. So it is, on a test piece broken statically after fatigue tests (Fig. 4.10b) that we find a brittle ring centred round an initial defect and developed progressively to the surface. Around that zone a classic progressive fatigue crack ring has begun, until the statically broken zone, characterised by a ductile appearance. This appearance is understandable because, during its development, the brittle white spot reaches the surface, freeing the hydrogen which continues to diffuse towards the exterior and thus suppresses its embrittling influence; cracking then continues normally by fatigue. This proves not only that in the case of cyclical stress, hydrogen can generate embrittlement under stress lower than the yield strength, but also that the resultant white spot can serve to initiate modification of fatigue performance. Hence for cyclically stressed constructions, the additional advantage of ensuring low hydrogen content in the weld metal.[17]

We shall only partially deal with cold cracking in this chapter which is devoted to development of the fusion zone, because in addition to the hydrogen involved, its mechanism involves hardening (of the weld metal or HAZ) and stresses. Here we shall only explain the process of passage of hydrogen towards the parent metal through the fusion boundary (for later study of the consequences). To this end, we need only consider on a longitudinal section of a weld bead (Fig. 4.11) the influence

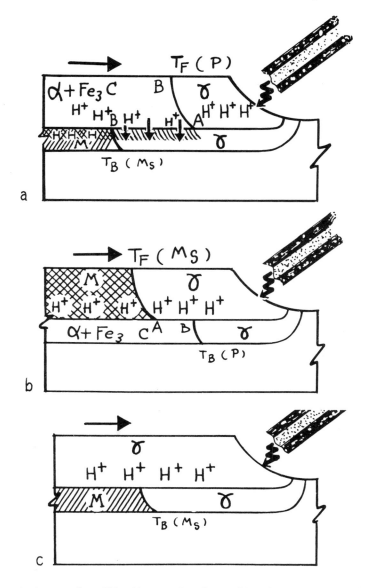

4.11 Longitudinal section of a weld bead being made: influence of the relative hardenability of the weld metal and parent metal on the passage of hydrogen from the first towards the second across the fusion boundary: a) Weld metal less hardening than the parent metal: isotherm T_F is in front of isotherm T_B; the hydrogen leaves the weld metal which has become ferritic and passes through the fusion boundary to the still austenitic parent metal where it is more soluble but less diffusible; on the passage of isotherm T_B, the martensitic transformation occurs in the heat affected zone in the presence of the hydrogen thus collected; b) Weld metal more hardening than the parent metal: isotherm T_F is behind isotherm T_B; the mechanism of hydrogen passing through the fusion boundary does not occur and the martensitic transformation takes place in the weld metal in the presence of hydrogen; c) Weld metal free from transformation point (austenitic filler): diffusion does not take place as the weld metal remains austenitic and the transformation on cooling takes place in the heat affected zone in the absence of hydrogen.

of the relative positions of the transformation isotherms of the melted metal T_F and parent T_B in the affected zone, depending on whether T_F is in front of or behind T_B.[18]

When the weld metal undergoes pearlitic transformation on cooling, therefore at relatively high temperature (case a), whilst the parent metal undergoes martensitic transformation, therefore from point M_S at relatively low temperature, isotherms T_F and F_B are offset in the direction shown in the figure. There is therefore a time gap AB during which the fusion boundary separates weld metal which has become ferrito-pearlitic, where the hydrogen has become less soluble (see Fig. 4.6), from an affected zone of the parent metal which is still austenitic and deprived of hydrogen whilst that gas is soluble there. The result is a diffusion mechanism from the weld metal towards the parent metal, encouraged both by the difference in solubility and concentration on either side of the fusion boundary. However, the hydrogen, which is less diffusible in austenite than in ferrite (Fig. 4.9), does not diffuse deeply in the HAZ where the arrival of isotherm T_B finds austenite loaded with concentrated hydrogen over a shallow depth. This is when martensitic transformation occurs in this austenite, hence the formation of a martensite which is already brittle and therefore susceptible to cold cracking, a phenomenon to which we shall return later. So this phenomenon affects the base metal on account of the hydrogen coming from the weld metal which, in this case, is free from cracking.

On the other hand (case b), if the weld metal is more hardening than the parent metal, it undergoes martensitic transformation at point A, later than the transformation of the parent metal at point B, especially if this latter is pearlitic; there is then no diffusion of hydrogen across the fusion boundary between B and A towards the parent metal, because the weld metal, still austenitic, may still retain this gas in solution. But when martensitic transformation of the weld metal occurs in A, it is martensite embrittled by hydrogen which forms, hence the possibility of cold cracking, but this time in the weld metal, whilst the HAZ is free of cracking.[19]

Finally, configuration c takes account of what occurs when, as often recommended, an austenitic filler is used to weld a hardening steel. Part of the success obtained by this prescription lies in the fact that the weld metal remains austenitic until completion of cooling, with no transformation which modifies hydrogen solubility. There is therefore no diffusion to the HAZ and consequently no embrittlement in that zone.[20] The plasticity of the deposited austenitic metal also contributes to this success.

Using the above explanations, we shall later in this book return to the methods of mastering the factors which cause cold cracking, namely hydrogen, martensite and stresses (see Chapter 8).

Hydrogen in welding of non-ferrous metals

Aluminium alloys provide an example of hydrogen's role vis-a-vis the formation of porosities of fused metal in MIG welding. The hydrogen source here is humidity adsorbed by product surfaces and wire skins. Broken down in the electric arc, this humidity forms hydrogen which, when released during solidification, generates porosities.[21]

In oxyacetylene welding of copper, the hydrogen present in the flame reducing zone plays a favourable role vis-a-vis the copper oxide and a sufficiently slow cooling permits the water vapour produced to be released. This is not the case with inert atmosphere welding; so it is necessary to use deoxidised copper as filler.

Influence of liquid medium: coatings and fluxes

General points

The liquid medium in question here is produced by fusion of coatings and fluxes; it accompanies the metal supplied by the electrode or wire in its transfer and covers the molten pool until it has solidified. Additional protection is provided during cooling by the slag supplied by the solidification of this liquid. In addition to their electrical (arc stabilisation, AC welding) and physical (joint formation, welding position) roles, coatings and fluxes play a metallurgical role, not only of protection but also reacting chemically during the time there is contact between molten metal and liquid slag. It is this latter role that we shall be examining in the following pages.

Before doing so, we need to make a general comment, resulting from the short duration of the process, which lasts only for the time taken for the arc to pass a given point. Because of this, no chemical reaction or exchange taking place achieves the state of equilibrium. The composition of coatings and fluxes must therefore be adjusted with a great deal of accuracy, care also being taken with granulometry and uniformity of the mixture of the necessary components. It is well understood that the corresponding products are not specified on the basis of their chemical composition (relatively secret in any case) but rather on the basis of the chemical composition of the metal produced or the mechanical properties of that metal. The coatings (electrodes and cored filler wires) together with powder fluxes are specified only on the basis of rather general classifications.

To make a more thorough study of the metallurgical role of coverings and fluxes, we can do no better than refer to the instructions of the relevant standards and comment on them.

Standards relative to covered electrodes and solid fluxes

From the metallurgical point of view, electrode coverings provide to varying degrees deoxidation of the metal[22] and, where applicable, transfer of the alloy elements into the molten pool (or even iron in the case of high yield electrodes) in order to give the desired properties to the deposited metal. Whether the steel is alloy or non-alloy, the standards concern only the type of covering and the characteristics of the deposited metal, submitted, depending on grades, to mechanical tests, chemical analysis or hydrogen dosing. The composition of the core is never specified.[23]

French standard A.81 309, directly derived from the corresponding ISO standard, provides symbols of seven types of covering, namely:

A: Acid (iron oxide)
AR: Acid (rutile)
B: Basic
C: Cellulose
O: Oxidising
R: Rutile (mean thickness coating)
RR: Rutile (thick coating)

Electrodes with type A, AR or O coating are essentially used for work on non-alloy steel requiring economic operation or particular bead shape conditions. The standard specified that the electrodes adapt badly to carbon or sulphur charged parent metal (hot cracking), and that A electrodes supply a low carbon content deposit. The hydrogen content of the deposited metal is not taken into consideration.

Basic electrodes (B) have a calcium carbonate or other basic carbonate based covering. They also contain fluorspar which encourages fluidity of the slag. Because of its basic character, this covering provides good protection, whether for alloy or non-alloy steel, and hence, good density of the weld metal and good mechanical properties. Because of the nature of the products of which they are made, basic coverings require particular care in storage, drying and handling to ensure the lowest possible hydrogen content.[24] It is only subject to these conditions that the use of these electrodes is recommended to avoid cold cracking in welding of steels likely to harden during the operation.

Cellulose electrodes (C) include a covering containing a large quantity of organic substances whose breakdown in the arc produces an abundance of protective gas, whilst there is little slag which is easily removed. The presence of hydrogen facilitates penetration and guarantees high fusion speed. Furthermore, it is possible to weld in all positions. This group of qualities means that this type of electrode is preferred for welding oil and gas pipes. But this use poses a problem because of the abundance of hydrogen in the weld metal which is not very compatible with the rapid welding of a hardening steel. Special techniques have to be employed to avoid cracking (see Chapter 8).

Type R or RR electrodes (the difference lies in the thickness of the coating) contain approximately 50% rutile (titanium oxide) to which can be added cellulose products (content not more than 15%). The beads obtained, of good appearance, are not deeply penetrated, the thick slag is often self-detachable. R type coverings are compatible with alloy elements,[25] but it should be remembered that the hydrogen content is relatively high, which makes the use of this type of electrode delicate for welding hardening steel.

There is still no standard concerning products involved in the make-up of cored filler wire electrodes for arc welding. By analogy with covered electrodes, we can expect standardisation of a rutile type and basic type. For this latter, the same drying and storage precautions need to be taken with regard to hydrogen as those recommended for basic electrodes.

For submerged-arc welding, French standard A81-319 lists five types of fluxes classified and symbolised on the basis of their principal mineral constituents. One or more additional symbols indicate, where necessary, the presence of metal additions. The following table concerns the main mineral constituents of the five types of flux selected in the standard:

Symbol	Principal constituents	Type of flux
MS	$MnO + SiO_2 > 60\%$ $CaO < 15\%$ $ZrO_2 < 5\%$	Manganese-silicate
CS	$CaO + MgO + SiO_2 > 60\%$ $CaO > 15\%$	Calcium-silicate
AR	$Al_2O_3 + TiO_2 > 45\%$	Alumina-rutile
AB	$Al_2O_3 + CaO + MgO + CaF_2 > 55\%$ $Al_2O_3 \geqslant 20\%$ CaF_2 (F total) $\leqslant 20\%$	Basic-alumina
FB	$CaO + MgO + MnO + CaF_2 > 50\%$ $SiO_2 \leqslant 20\%$ CaF_2 (F total) $\geqslant 15\%$	Basic-fluoride

With regard to these fluxes, standard A81.319 gives comments from which extracts are given below:

MS: Gain in Mn and Si in the weld metal if the electrode wire has a low Mn content.

CS: Same comment as for silicon, if the Si content of the wire is low. Possibility of using high intensities.

AR: Usually agglomerated fluxes,[26] permitting the use of the 'multi-wire' technique and high speed welding.

AB: Generally agglomerated, intermediate between AB and FB.

FB: Generally agglomerated, low SiO_2 content, designed to reduce transfer of silicon into the fused metal and obtain high levels of notch toughness in multipass welds.

Having specified these flux characteristics, we must remember that the final composition and, consequently, the properties of the weld metal are dependent on the 'wire-flux couple' used in the preparation. The wires themselves are specified and symbolised on the basis of their chemical composition.

Finally, it should be remembered that the problems posed by hydrogen in the fused weld metal are not limited to arc welding with covered electrodes, at least in the case of welding high strength steel. Hence the precautions in drying and storing fluxes, in particular agglomerated fluxes, which may be hygroscopic.

Influence of solid medium: dilution and its consequences

Summary - dilution

The solid parent metal on which the molten pool rests and develops constitutes, after the surrounding gas and liquid mediums, the third factor involved in the determination of the chemical composition of the weld metal, by way of dilution. Dilution has already been defined as being the proportion in which the parent metal, by its own melting contributes to the development of the weld metal zone; a simple means of evaluating dilution by macrography has also been given (see Chapter 3).

The dilution value obviously depends on the welding process and, for a given process, on the procedure, in this case the preparation, the number of runs and welding energy. For example, for arc welding with covered electrodes or bare wire with gaseous fluxes, or even cored filler wire with or without gas, dilution reaches a maximum of around 20% for the first run of a butt weld on a V groove (Fig. 4.12a), whilst it may reach 80% or more for a submerged-arc weld on straight edges in one or two runs[27] (Fig. 4.12b); in actual fact, these values signify that, in the first case, the weld metal zone is mainly constituted by the added metal, whilst in the second case, it is the parent metal which provides the major contribution. We shall see later the consequences of this comment.

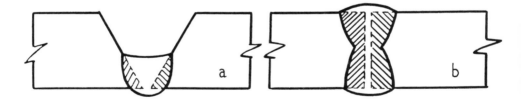

4.12 Comparison of dilutions, at equal thickness, for a first arc weld pass (a) and for a two-pass submerged-arc butt weld (b).

At the limits of this definition, we find on one side brazing and braze welding, on the other side spot or EB welding. For brazing and braze welding dilution is nil, at least on the macroscopic level, because the entire fusion zone is provided by the filler metal.[28] For spot welding, it is equal to 100% because it is the parent metal which supplies the whole of the fusion zone. Such is also the case for EB welding, except in the relatively rare case where a filler wire is used.

As consequences of dilution we shall examine below its role vis-a-vis the properties of the weld metal and, more generally, weldability, the need to take it into account for the specification of filler products and finally its involvement in heterogeneous welds.

Properties of the weld metal zone - weldability

The participation of the parent metal in the formation of the weld metal zone, depending on welding process, implies that the properties of the weld metal zone are different depending on the value of the corresponding dilution: for example, the longitudinal joints of oil or gas pipes, generally carried out by automatic submerged-arc welding, are much more influenced by the composition of the parent metal than are butt joints, whether they are carried out by manual covered electrode welding or by automatic MAG welding. It is all the more so for spot or EB welds, already different as regards their heat cycle, in which the molten metal is furthermore entirely provided by the parent metal. So we have to consider that the standards concerning weldable steels[29] correspond only (for chemical composition range) to a mean deduced from their behaviour in various applications and different welding processes. It is often advisable to complement their use with individual specifications taking account of the characteristics of those applications and processes.[30]

Specification of filler products

The moderate dilution, indicated above, which characterises the majority of arc welding processes justifies the position taken by the standards relative to filler products, namely that the chemical or mechanical characteristics are determined on the undiluted deposited metal provided by these products. Hence the practice of tests on a 'mould' for which French Standard A81-302, relative to testing of covered electrodes, specifies the procedure (Fig. 4.13a).

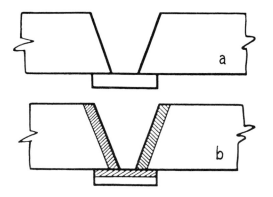

4.13 Preparation specified by standard A 81-302 for depositing weld metal for mechanical tests. The distance between edges is designed to limit dilution in the central part: a) Mould for non-alloy or low alloy steel welding; b) Mould for alloy steels: prior surfacing of edges and backing strip.

This mould is the same type for filler products (bare or cored filler wires used in processes such as MIG or MAG). Its design, together with the arrangement of the runs and position of samples taken, reduce to a minimum the dilution coming from the steel used for the mould. It should also be stated that, for non-alloy or low alloy steels, the mould is specified without particular reference to the product tested, whilst for stainless and refractory steels, it is either constituted by a grade of steel identical to that of the metal deposited by the product tested, or by non-alloy steel, the chamfer and support being then surfaced with the filler product tested (Fig. 4.13b). Similar precautions are taken for constituting the deposit intended for chemical analysis or the hardness test.

So, subject to the conditions just indicated, we can characterise filler products used with a relatively low dilution by tests carried out on the deposited metal they provide.[31] It is quite different in the case of submerged-arc welding which, as we have seen, can be used either like other arc welding processes, in multipass welding, or in welding with one or two runs (one on the face side, one on the reverse side). In this latter case, where dilution reaches high values, tests on the undiluted deposit metal would make no sense. This is why the normalisation relative to the wire-flux couple only provides for tests where the product is used in multipass welding, leaving aside the case of high dilution welding, which can only be dealt with by individual specifications relative to each application, since account must be taken of the nature of the parent metal.

Heterogeneous welds (Cr/Ni alloy steels)

The problems posed by dilution become complicated when we talk about assemblies which are called heterogeneous, i.e. involving as parent metal or filler metal materials of deliberately different chemical composition.[32] In the weld metal zone, this difference results in a mixture, in variable proportions, of the constituent elements and the main problem which arises is that of compatibility of those elements, which must constitute in the weld metal zone a satisfactory alloy both at the time of welding and in service.

① $B_1 = B_2$ $A \neq B_1$ and B_2

② $B_1 \neq B_2$ $A = B_1$ or B_2

③ $B_1 \neq B_2$ $A \neq B_1$ and B_2

4.14 Practical cases of chemical composition heterogeneity concerning parent metal B and filler metal A.

Three possible instances of deliberate heterogeneity of welded assemblies involving Cr/Ni steels can be shown in diagrammatic form by Fig. 4.14, which can be illustrated by the examples below, relative to assemblies involving non-alloy or alloy steels, which does not mean to say that there are no instances of heterogeneous assemblies in the field of non-ferrous metals.

Case 1 (parts for assembly B of the same composition, filler metal A different) can be illustrated by the welding of hardening steel by means of an austenitic or austeno-ferritic Cr-Ni or Cr-Ni-Mo type alloy steel filler. The use of such filler makes it possible to prevent the risk of cold cracking which characterises welding of this type of steel; we use here the particular behaviour of hydrogen in the austenitic molten metal which has already been described earlier in this book (see Fig. 4.11c).

Case 2 (parts for assembly of different composition, filler identical to one of them) is that of welding pipes constituted on the one hand by 'black' steel (non-alloy or low alloy steel), and on the other hand by austenitic steel.

Case 3 (parts of different compositions, filler different from both) is that of making a weld bead affecting simultaneously the cladding and the support steel of a non-alloy steel sheet plated with 17% Cr ferritic steel, the filler being provided by Cr-Ni austenitic steel.

These applications, and others of the same kind, (in particular concerning surfacing) have led to the search for a representation making it possible to forecast, for one of the above three configurations, the constitution of the weld metal F (i.e. diluted) depending on the chemical compositions of the parent metal(s) B together with filler metal A (undiluted) and, of course, on the dilution. To that end, we must be able to express each chemical composition in question by a couple of representative values – Cx and Cy, which make it possible, by means of a simple geometrical construction, to place the representative points of metals B_1, B_2 and A on a two dimensional graph. It is then possible, in the three above-mentioned cases, to make a simple graphic construction to find the representative point F of the molten metal for each combination, provided that the characteristic dilution of the processes and procedures involved is known (Fig. 4.15).

By using this construction, we can for Cr and Ni based alloy steels, as proposed by Schaeffler[*], go from the chemical composition to the constitution of the corresponding alloy, (after Schaeffler, De Long[**] introduced the nitrogen effect) by transposing, for the weld metal, a representation already proposed for cast parts under the name of Maurer's[***] diagram.

---

[*] A L SCHAEFFLER: Constitution diagram for stainless steel weld metal – Metal Progress, November 1949, pp 680-680b

[**] W T DELONG: Measurement and calculation of ferrite in stainless steel weld metal – Welding Research, supplement to the Welding Journal, November 1956, pp 521 s – 528 s

[***] E MAURER, B STRAUSS: High nickel chrome alloy steels used as stainless steels. Kruppsche Monatsch, 1920, pp 120-146 – Welding Metallurgy quoted by D SEFERIAN, 1959, pp 275-276

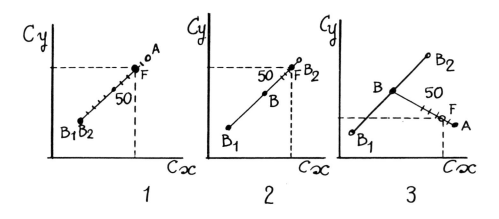

4.15 Graphic determination of the chemical composition of weld metal F in relation to dilution in the three examples in Fig. 4.14. Dilution assumed equal at 20%. Example 1: graduation from 0 to 100% on line A – B_1 B_2. Point F is 20% from point A. Example 2: false composition B is marked (in the middle of straight line B_1 B_2). Point F is then 20% of B B_2 from B_2 as the filler has the same composition as B_2. Example 3: false composition B is marked as above, representative point A of the metal deposited is located and point F is found between A and B as in the first example.

In this type of representation,[33] values Cx and Cy are calculated from the content of elements which, like chromium, are said to be 'alphagenic' (Cr, Mo, Si, Nb) or, like nickel, are 'gammagenic' (Ni, C . . .), depending on whether these elements favour the formation of ferrite or austenite. Hence the equivalent Cr (Cr_{eq}) and equivalent Ni (Ni_{eq}) terms which appear respectively in abscissa and ordinate of the representation in question, shown in Fig. 4.16, which indicates the formulae used and where the limits of existence of ferrite α, austenite γ, and martensite M are traced, alone or in association with each other. To forecast the constitution of the weld metal in one of the three cases of heterogeneous welding mentioned above, it is necessary to calculate the chrome and nickel equivalents of the parent metals B_1 and B_2 and of the filler metal A (undiluted), and, provided the dilution is known, produce one of the three geometric constructions of Fig. 4.15 on the representation of Fig. 4.16.

However, this representation should not be considered as quantitatively accurate, in particular with regard to the ferrite content of the fused metal in the austeno-ferritic field. Indeed, to locate the representative point of the fusion metal, we assume that that fusion metal is homogeneous, which is not the case, especially, as we have seen earlier, in the vicinity of the fusion boundary. Also, the constitution indicated concerns the as-solidified weld metal of one welding run; it is not applicable in the case of several runs or especially after heat treatment. Be that as it may, this method provides a mean indication, useful for everything concerning the estimate of weld metal zone behaviour (susceptibility to corrosion, magnetism, hot cracking and mechanical properties). Evaluation of the ferrite content is debatable.

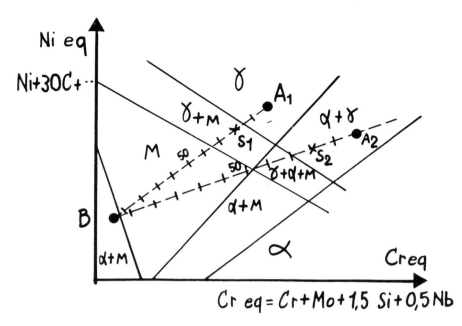

4.16 Schaeffler representation relative to the constitution of the weld metal in Cr-Ni steel welds. Example of application to example 1 of preceding figure: if the parent metal B is welded with austenitic filler A_1, a 20% dilution results in austeno-martensitic weld metal S_1. For the same dilution, austeno-ferritic filler metal provides weld metal S_2 which is also austeno-ferritic.

Variations in chemical composition at fusion boundary

The fusion boundary in a fusion weld is the surface produced by the displacement of the side part of the molten pool, which separates the unmelted parent metal and the liquid metal of the molten pool. It is on this surface that solidification begins, under conditions which will be studied later (Chapter 5). In welding processes with pressure and momentary production of a liquid phase, such as flash welding, the fusion zone unites two parts which have remained in the solid state, but which during the operation have generated a liquid phase which the pressure has eliminated.

In a pure metal, the production of a liquid phase by the parent metal obviously causes no modification of chemical composition. On the other hand, with an alloy, this production arises as soon as the temperature reaches that of the 'solidus', from which there is coexistence of two liquid and solid phases, which is the case for example for the steel in the upper section of the Fe-C diagram, represented on Fig. 4.17 in its simplified form, i.e. not taking account of the formation of delta ferrite. In the solidus, i.e. near the molten pool, the parent metal in which the carbon content is $C_s\%$, produces a liquid phase on which the initial C_L carbon content is greater

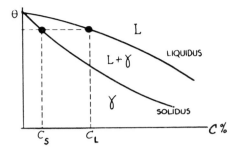

4.17 Decarburisation mechanism at fusion boundary in steel welding.

than C_S when the temperature reaches that of the solidus. At the time of welding, the liquid phase emitted by the parent metal is absorbed by the advancing molten pool, but, on the side of that molten pool, which produced the fusion boundary, it is a liquid richer in carbon than itself that the parent metal must provide along the length of the fusion boundary.[34]

This happens by diffusion, which results in the existence of a narrow decarburised band in the vicinity of that zone, which is also found in fusion welds as well as flash welds, since the elimination of the fused metal generates the same phenomenon.[35]

This process of steel decarburisation, due to the temperature gradient in the fusion zone, may possibly be reinforced by the process resulting from diffusion across the fusion zone, in the case of heterogeneous welds made with a filler in which the carbon content is low in relation to the parent metal. This carbon transfer by diffusion may continue during subsequent heat treatments, in particular in the case of non-alloy or low alloy steel welds with Cr-Ni filler metal. Such treatments must be carefully considered.

The behaviour just described concerns the case of complete solubility in the solid state. If the alloy in question, account taken of its alloy or impurity contents, includes solid phases with a low melting point, or liable to form a eutectic, also with a low melting point, the phenomenon of intergranular liquation, i.e. the formation then solidification of a liquid phase at grain boundaries, is liable to occur at the fusion boundary. It may produce a mechanical failure ($Cu-Cu_2O$ eutectic formation in copper welding) or chemical failure ('incisive' corrosion of stainless steel by excess titanium). It can even happen that the intergranular liquation causes hot cracking (effect of sulphur on nickel and its alloys, of silicon in Cr-Ni austenitic steel welding), especially if the liquation in the parent metal is prolonged in the intergranular segregation which can affect the melted metal during its solidification[36] (Fig. 4.18). Such a configuration is encouraged by the epitaxial nature of solidification, which has already been mentioned and to which we shall return during our study of solidification.

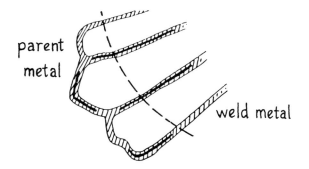

4.18 Intergranular liquation in the parent metal and intergranular segregation in the weld metal: the epitaxy at the fusion boundary favours the passage of hot cracks from one metal to the other.

Chapter 4 **Comments**

1 The same applies for magnesium, although eminently volatile, whilst the loss of magnesium is significant in Al-Mg alloys.

2 Hygienists prefer to use the term 'dust', 'fumes' being defined as being a mixture of dust and gas.

3 This is the reason why we recommend the use of an Al-Mg type filler for welding these alloys.

4 Cadmium fumes are highly toxic. It is strongly recommended that welding on cadmium-coated parts is not done unless special precautions are taken.

5 At the same time as it causes porosities to disappear, hot rolling produces lengthening and flattening of malleable inclusions, like manganese sulphur. The corresponding morphology plays an important role vis-a-vis mechanical characteristics and the cracking phenomenon called 'lamellar tearing', already mentioned in Chapter 2.

6 It is this difference in homogeneity between rimming and killed products which, in standards relating to steels, is the reason for the idea of analysis at the time of casting and on the product.

7 Such an arrangement is also unfortunate from the point of view of the risk of lamellar tearing, whether the steel is rimming or killed.

8 Nitrogen, as well as oxygen, spoils the mechanical characteristics of unprotected steel welds. But the precautions taken against the influence of oxygen provide sufficient protection against nitrogen, which is present in steel welds at contents not leading to the essential risk, namely susceptibility to ageing, which remains moderate.

9 Despite the presence of hydrogen in the gases of the oxyacetylene flame, there is little involvement of the weld metal fixation mechanism because the dissociation of $H+$ ions is low at molten pool temperature and furthermore cooling is relatively slow.

10 It should be noted that Sievert's curve gives information on the solubility of hydrogen in the equilibrium state. It can therefore only be used qualitatively in the case in point, where we are far from that state.

11 It is in this way that French Standard A81.301 specifies a 2 hour treatment at 250 °C for tensile test piece blanks.

12 More often than not it is possible to reach 450 to 500 °C for basic electrodes without damage.

13 Dislocations are imperfections in the position of the atoms constituting the crystal network. They are liable to move under the effect of a stress.

14 These are macroscopic inclusions, as distinguished on fractures, with the naked eye or low magnification if need be.

15 Hence the practice, already mentioned, of a degassing treatment concerning tensile test piece blanks on deposited metal and not those of notch toughness test blanks.

16 The same mechanism applies, it seems, to the role played by parent metal inclusions with regard to cold cracking, which, paradoxically, would be encouraged by a reduction in impurity content.

17 It goes without saying that the stress relieving heat treatment can only be beneficial in this respect, but it is not always materially possible.

18 To clarify the position, the gap between T_F and T_B can be of the order of about 10 seconds if T_F corresponds to a pearlitic transformation and T_B to a martensitic transformation, as is the case for Fig. 4.11a.

19 Between a and b, it is possible to meet a case where, weld metal and parent metal having the same hardenability, both are affected by embrittlement and therefore cold cracking.

20 It is however necessary to avoid excessive dilution which would render the weld metal zone more or less martensitic and thus susceptible to embrittlement.

21 We must also point out the phenomenon called 'swelling' which occurs in the parent metal, near to the fusion boundary. But the hydrogen in question here comes from the conditions of preparation of the alloy, in which the hydrogen is concentrated around the inclusions.

22 To this end, all types of coating contain a certain proportion of ferromanganese.

23 A special symbol (x) is however reserved for 'synthetic' electrodes which deposit a highly alloyed metal from a mild steel core.

24 Standard A81.309 specifies the symbol H for a hydrogen content of up to 10 ml/100 g and BH for a content up to 5 ml/100 g (determination in accordance with French Standard A81-305.

25 There are rutile covered electrodes for welding Cr-Ni stainless steels.

26 In effect, with regard to manufacturing methods there are two families of fluxes, one fused, the other agglomerated or sintered with the aid of a suitable binding agent.

27 For multiple run welding on chamfered edges, submerged-arc welding belongs to case a.

28 In reality, even if it is not apparent on the macroscopic scale, a certain amount of dilution, however low it may be, is involved in the brazing process by diffusion of the atoms of the substrate metal which become diluted in the molten metal.

29 We prefer the expression 'steels for welded constructions'.

30 This is also true for the heat affected zone.

31 It should however be noted that such tests, the results of which are associated with an accurate process procedure relative to the constitution of the mould, take no account of the performance of the actual joints. Hence the relevance of the tests carried out directly on those joints, to qualify process procedures (see Chapter 10).

32 In reality there is practically no completely homogeneous welding from the chemical point of view, if only because of the characteristics of weld metal preparation.

33 In reality, the term 'diagram' is incorrect, because neither temperature nor state of equilibrium are involved. This is a representation showing the influence of chemical composition on the constitution of the as-solidified weld metal.

34 Contrary to what could be believed owing to the period between liquidus and solidus generated by welding, the phenomenon called burning (well-known in iron and steel metallurgy) is not involved. This is due to the temperature gradient which does exist in the fusion boundary but not in a mass of steel uniformly heated in the same temperature range.

35 The decarburised band can be seen clearly only in the heat treated condition. In the as-welded state, it is hidden by the combined effect of overheating and cooling transformations.

36 Sulphur is often cited as an element favouring hot cracking in steel welding. In fact, the reduction to low sulphur content values in modern steels means that this phenomenon is now exceptional.

5 Solidification of the weld metal zone

Characteristics of solidification

After our study in Chapter 4 of how the weld metal develops during the welding operation, we propose to study how it solidifies to produce the weld metal zone.[1]

Having mentioned in the introduction to this book the comparison made by Portevin between welding and casting, we need (as we did for formation) to emphasise the differences between the solidification of weld metal and that of moulded items, if only to extract the majority of similarities.

In an ingot or a moulded part, the initial situation is one of an immobile liquid mass,[2] except in centrifuging, at an even temperature at the heart of which a temperature gradient is only established, from the mould, from the commencement of cooling by heat exchange between the outside of the mould and the medium surrounding it. Solidification starts on the mould walls, without fusing the moulding with the mould and progresses with the isotherms towards the interior of the part. In fusion welding, it is the displacement of the liquid mass that is represented by the molten pool which generates solidification jointly with the parent metal; as in formation, solidification of the fusion zone occurs step by step and the process of solidification is regulated by the speed of welding. Also, at the liquid-solid interface during displacement, there is an intense temperature gradient, both in front of (in the solid phase) and behind that interface.

The nearest example to the casting operation is, except for scale, that of spot welding which will therefore be looked at separately. We can say now that, although in this case we have cooling of a macroscopically immobile liquid phase, subsequent solidification and cooling take place under pressure.

After describing the molten pool, we shall study below the crystallography and physico-chemistry of solidification, to arrive at the practical consequences of the considerations thus developed.

Study of the molten pool

General case

If we leave aside EB and spot welding, the molten pool (Fig. 5.1) may be defined as being, at a given moment, constituted by a certain volume of displaced metal in the liquid state produced by the melting of the parent metal and possibly fed by a filler. It is limited in front by the parent metal fusion isothermal surface and behind by the liquid/solid interface resulting from its advance. Depending on the welding process, its surface is surmounted by the protective gas or liquid slag arising from the flux or coating.[3] This surface is not even because it is subject to the impact of the heat source (flame, arc, plasma or laser) which has a not insignificant dynamic effect and also that of the temperature gradient which creates convection movement. The electromagnetic forces generated by the arc also play a role in the molten pool. Finally, the movement of the heat source results in movement of the molten pool, which retains its shape if the temperatures, together with the filler metal regime, are maintained in the quasi-stationary state.

Thus, the advance of the heat source and as a consequence of the molten pool generates movement in the weld metal from the front, where it forms, towards the rear where it solidifies, both laterally and depthwise (Fig. 5.2). Furthermore, the more or less elongated shape of the molten pool, depending on welding process and speed of advancement, has the result that all dl advancement of the heat source (Fig. 5.3) results in the liquation of a certain volume of metal dV, to which corresponds the solidification of the same volume behind. But this volume occupies a smaller thickness δh than dl, so that it is possible on the longitudinal section of a weld bead, or on its surface, to detect the pattern amplified by any anomaly which has occurred at the front of the molten pool, associated with a variation in welding speed or chemical composition.[4] We find also, on such a section or on the surface of weld beads the

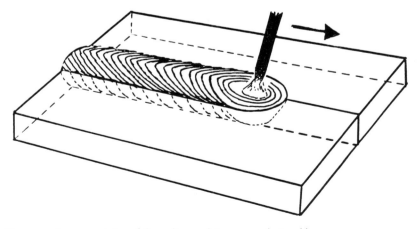

5.1 Diagrammatic representation of the molten pool, in a square butt weld.

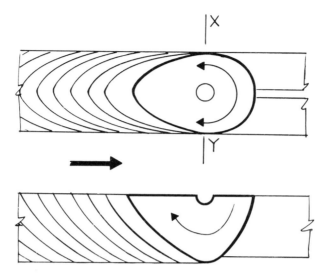

5.2 Plan view and longitudinal section of the molten pool and movements of the liquid metal inside the pool. The striations visible on the surface illustrate the progression of the molten pool.

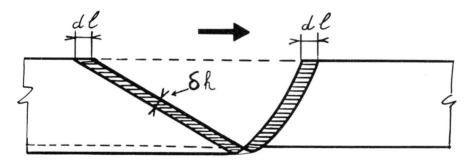

5.3 Longitudinal section view of the effect of advance dl of an elongated molten pool: thickness δh of the corresponding solidified layer becomes lower the more the pool is elongated.

pattern of variations in solidification speed caused by supercooling, or even by the frequency of the welding current. This explains the ribs that we see on the surface, more or less elongated depending on welding speed and process used.

Electron beam and spot welding

We have already indicated that in EB welding, the molten pool is not comparable, as regards its form and process, with the molten pool of other fusion processes: whether the joint is a flat butt (and the beam vertical) or whether it is horizontal in a vertical plane (and the beam horizontal), the molten pool is, continually, made up of a cylindrical cavity hollowed out by the electrons and surrounded by molten metal covering the walls. As is usual in welding, the pro-

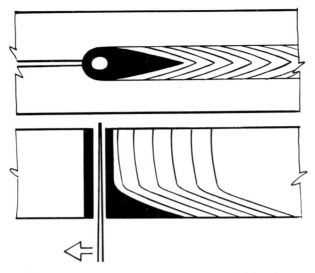

5.4 Electron beam welding: formation of weld metal zone (square butt weld) by advancement of a cavity generated by the beam, lined by a liquid sleeve (in black on the longitudinal section and plan view).

gress of this cavity produces melting of the metal in front and the bond in the liquid state at the rear (Fig. 5.4), with a tail resulting from the rapid advance of the beam and loss of energy in its thickness. Like Arata* did by high-speed radiocinematography, we can show the form and progress of the molten pool by using the contamination method mentioned in note 4. Due to this process, the stirring up of the liquid phase is less intense than for standard fusion welding processes. Account should be taken of this in the case of heterogeneous welds because, after solidification, we see a clear dissymmetry, each side of the fusion zone having a chemical composition close to that of the parent metal from which it came, for example in the case of a Cr-Ni austenitic steel weld on Cr martensitic steel (Fig. 5.5). The hardness relationship reveals the macroscopic heterogeneity of the fusion zone.

Although it is immobile, the molten pool of a resistance spot weld (Fig. 5.6) shows a noticeable characteristic compared with an ingot or a moulded item.[5] Indeed, heating commences under the effect of the inter-part contact resistance, but, as soon as a metal bridge is established (by pressure welding), the Joule effect occurs at that bridge and produces a localised fusion zone there; this zone enlarges with the duration of the welding current. The fusion zone is therefore formed by centrifugal displacement of isotherms. As soon as the welding current is cut, cooling by conductivity in the part and towards the electrodes, which remain in contact, reverses the isotherm displacement, and solidification progresses towards the interior, as with a mould but obviously much more quickly (phases a, b, and c of Fig. 5.6).

* ARATA: What happens in high energy density beam welding and cutting – 1980

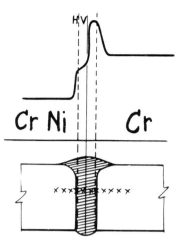

5.5 Electron beam welding: demonstration of incomplete homogenisation of the molten pool, by a hardness relationship in the case of a heterogeneous weld (A austenitic steel, M martensitic steel).

Having thus described the molten pool, we can now study the initiation and development of solidification, from the fusion boundary where initial crystal orientation is established, then along the solidification path which is imposed on their growth by displacement of the isotherms.

Crystallography of solidification

Orientation of crystals in the fusion boundary: epitaxy

For a given position of the molten pool, account taken of its form and displacement, solidification starts on the parent metal/molten metal interface, at the spot where its transverse section is maximal (XY on Fig. 5.2), the section we shall find again on the macrograph of the completed weld. On this interface, the displacement of which constitutes the fusion boundary, the crystallisation of the solidifying metal occurs in accordance with the epitaxial mode (mentioned in Fig. 5.7, which repeats Fig. 1.6), in which the solid crystals forming adopt the orientation of the crystals of the substrate on which they lie. Micrographic examination is enough to confirm this characteristic: in effect, each grain being an edifice formed from identically orientated crystals, the grain boundaries constitute boundaries between differently orientated edifices. If therefore the solidifying weld metal adopts at every point the orientation of the underlying substrate, we will see the grain boundaries of the substrate cross the fusion boundary and continue without a break their extension in the boundaries of the grains of the solidified metal.[6] Such is the starting position of the solidification

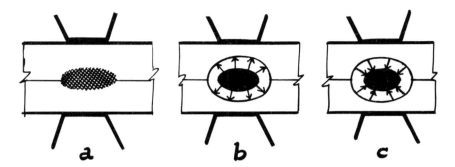

5.6 Centrifugal formation from the weld metal zone of a resistance spot weld: a) Pressure, b) Fusion, c) Solidification.

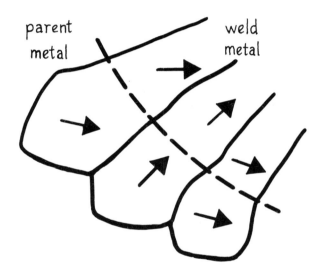

5.7 Epitaxial solidification in the weld metal zone. This zone, delineated by broken line on the figure, is not visible in the case of pure metal.

grains, but we shall see below that another phenomenon called selective growth will soon modify that position.

An important consequence of the epitaxy at the fusion boundary lies in the fact that the size of the solidification grains is, at the start, determined by the size of the grains in the substrate adjoining that zone. It is here that the overheating phenomenon occurs, in a variable manner depending on the welding process, so that overheating of the substrate and the more or less coarse nature of the solidification grain are connected. Consequently, particularly in steels, any action tending to limit grain enlargement in the parent metal (process procedure, chemical composition) results in a beneficial effect with regard to the solidification grains in the weld metal zone.

Naturally, the epitaxial nature of fused metal solidification, which has just been described in relation to the parent metal, concerns also orientation at the interface between two successive runs, provided that the heating due to a run does not cause a change in the constitution of the preceding run on which it rests. Thus epitaxy is obvious, for example on multipass aluminium or aluminium alloy welds, or even austenitic steel (Fig. 5.8), because these materials are devoid of solid phase transformation points.

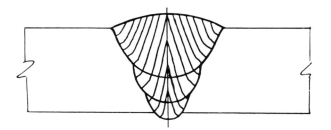

5.8 Epitaxial solidification of successive weld passes of a pure metal or an alloy free from transformation point: maintenance of the solidification structure across successive fusion boundaries.

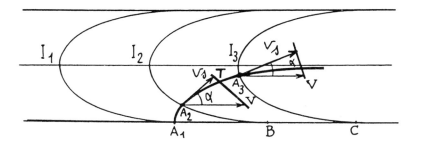

5.9 Solidification path generated by movement of the molten pool: V welding speed (constant), V_s imposed solidification speed (increasing along the length of the path, with $V_s = V \cos \alpha$).

Solidification path –solidification speed

From the fusion boundary, where the crystalline orientation of each grain is determined, solidification progresses as the molten pool progresses, grains tending to extend along paths at right angles to the isotherms, in accordance with the laws of solidification. So, for the successive positions $I_1 \, I_2 \, I_3$, etc, of the isothermic surface bounding the rear of the molten pool, we can define (Fig. 5.9) a solidification path $A_1 \, A_2 \, A_3$, etc, which starts at A_1 and curves in the direction of the displacement of the molten pool until it connects with the centre line. It is along these paths, which are warped curves, that the solidification structure is created, under conditions given in detail later.

But we must first state that the solidification speed V_s imposed on the weld metal is variable along this path, from a very low value or nil at the start (at A_1) to a maximum value in the centre line of the bead, where V_s becomes equal to the welding speed. Indeed, as solidification tends to progress perpendicular to the isotherms, speed V_s imposed on it is determined at each point by the projection of the welding vector speed V on the normal isotherm speed. We also see that the vector $V_s = V \cos \alpha$ rises from a value of nil at A_1 at the start to a maximum value equal to V at the centre of the joint where α is nil because solidification is progressing in the same direction as the molten pool.

Also, in addition to this continuous variation in solidification speed resulting from the shape and progress of the molten pool, we can see periodic or pseudo-periodic variations, associated with the heat source (including the frequency of the feed current) with the mode of transfer of filler metal and also, possibly, in the supercooling in front of the solidification front, or even, simply an irregularity in the progress of the molten pool, of purely mechanical origin, with automatic welding. These variations appear all along the weld bead in their consequences with regard to crystallisation. We find traces in the ribs on the surface and on longitudinal sections of weld beads.

However, to understand the process of solidification along the paths imposed by the displacement of the molten pool, we must follow the application of the laws of solidification to explain the phenomenon of selective growth which means that, for industrial metals or alloys, some grain grow at the expense of others or appear in their place. In the case of a pure metal, a crystal originating in the liquid phase is liable to develop at the same speed in all directions permitted by its crystal structure. The orientation imposed by the epitaxy at the liquid/solid interface has no influence on the growth of grains which, from their origin in the fusion boundary (Fig. 5.7) grow unchecked along the solidification path. On the other hand, the solidification crystals of industrial metals or alloys are characterised by a 'privileged growth direction'[7] according to which speed of growth is maximal. The result is that grains whose privileged growth direction coincides with the solidification path T (Fig. 5.10) enjoy favourable growth conditions at the expense of neighbouring grains whose chances of growth are less because they lie in a different orientation.

This growth, which is called selective, results in the disappearance of less favourably orientated grains and therefore a reduction in the number of grains with an increase in their size compared with their initial size (Fig. 5.11a). If the solidification path itself curves sufficiently, these grains lose their privilege to be replaced by others, from better orientated nuclei[8] (Fig. 5.11b).

In any case, for industrial metals or alloys, there is for each one an orientated solidification speed limit of which the compatibility with the speed imposed by the movement of the molten pool creates, depending on processes and procedures, the various solidification structure configurations; these configurations have been particularly well described by Matsuda*.

---

* F MATSUDA, T HASHIMOTO, T SENDA: Fundamental investigations on solidification structure in weld metal. Trans. of National Research Inst. for Metal, 1969, Vol. 11, no. 1, pp 43-58

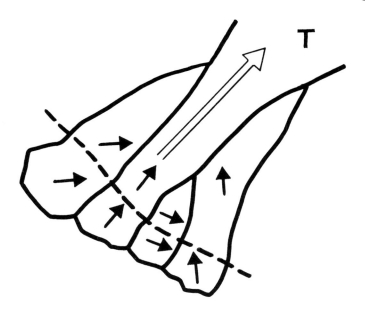

5.10 Selective growth mechanism at fusion boundary. The weld metal grains whose privileged growth direction coincide with the solidification path develop at the expense of the others.

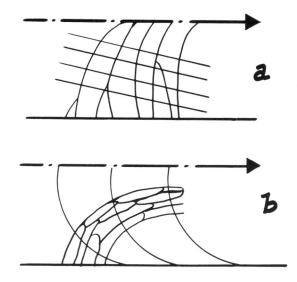

5.11 Solidification grains arising from the selective growth (plan half view): a) Slightly incurved solidification path (elongated molten pool); the solidification grains tend to reach the bead axis; b) Incurved solidification path (rounded molten pool); replacement of grains by grains better orientated along the path.

Applications to the macrographic aspect of the weld metal zone

Figure 5.12 gathers together the various macrographic aspects which can be revealed in the weld metal zone of fusion welds with displacement of the heat source[9] on sections cut parallel to the surfaces of assembled products (which is the equivalent of a plan view) and on transverse sections. These aspects depend both on the heat characteristics of the process in question applied to the material being studied, and on the speed of solidification of that material, compared with speed V imposed along the length of the solidification path.

Finally, it should be noted that, in the case of steel, a macrograph does not reveal only the solidification structure, but the structure resulting from the transformation on cooling which follows solidification. However, the latter is influenced by the former.

In case a, which is that of a relatively slow weld, the molten pool shows little elongation, hence the form of the solidification paths which are, to start with, perpendicular to the fusion boundary and curve progressively in the direction of the displacement of the

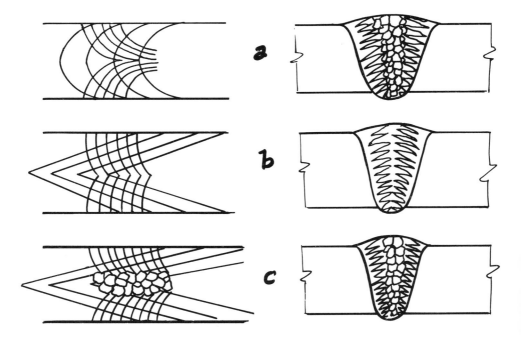

5.12 Principal aspects of the solidification structure, appearing on transverse sections and plan views: a) Slow welding, rounded molten pool: grain growth continues as far as the weld axis towards which the grains curve, perpendicular to the transverse section; b) Rapid welding, transverse grain development; the grains converge towards the central part, perceptibly in the transverse section direction; c) Rapid welding, speed too high in the central part compared with the speed of solidification; hence equi-axial solidification in that spot. The transverse section takes account of this change.

molten pool. If the speed imposed along these paths is, as far as the central section of the fusion zone, less than the solidification limit speed of the material in question, the grains (or strings of grains) grow without hindrance as far as the central part. The result of this is that, on the transverse section, we see, all around the fusion boundary, sections of grains of elongated shape, due to the fact that the solidification path makes a small angle with the section plane. On the other hand, in the central part, we find sections of grains of any shape[10] because the plane of the section is roughly perpendicular to the solidification path.

Case b takes account of a slightly different situation in which the elongated shape of the molten pool, associated with rapid welding, goes hand in hand with a quasi-transverse arrangement of the solidification paths. If, by its speed of solidification, the material in question can see its grains build up as far as the central part, we get the plan view and transverse section illustrated in Fig. 5.12b. The grains arising from the fusion boundary grow symmetrically to join up in the central part, so that, in that area, there is no longer a zone where the grains have developed perpendicular to the section plane as was previously the case. Hence the characteristic appearance of the transverse macrograph.

Configuration c corresponds to a case of rapid welding, in which the molten pool is elongated, as before, affecting this time a material in which the solidification limit speed makes impossible, at a certain place, the continuation of solidification orientated in accordance with the solidification path. Then crystallisation takes on another aspect and continues in an equi-axial manner, i.e. in the privileged growth direction, from nuclei of any orientation which appear in the liquid phase which, by supercooling, is slow to solidify. Then we obtain the macrographic aspects, in plan and transverse section, illustrated by Fig. 5.12c, aspects which should not be confused with those of case a, because here we have, in the central part, a structure which is equi-axial and not of perpendicular orientation to the section.

Naturally, the appearance of one of these three types of solidification macrostructure is a function of the welding procedure used for a given material and welding process. For example, in the case of vertical electroslag welding, it is possible to obtain, depending on the value of the voltage and welding current, either a molten pool and consequently practically flat isotherm surfaces, which encourage the development of a solidification grain curving in very rapidly to grow vertically (Fig. 5.13a), or a molten pool and then hollow isotherm surfaces, which generate solidification paths which are more transverse the more sunken is the molten pool (Fig. 5.13b). Hence the macrographic appearances obtained in each of these two cases.[11] With regard to the transverse macrograph illustrating case a, the apparent difference in the shape of the grains between the periphery and the centre lies both in the different orientation of the section examined in relation to the solidification paths, and in the already mentioned influence of the size of the grains in the parent metal at the fusion boundary.

With regard to resistance spot welds, the solidification paths are generated by the centripetal displacement of surface isotherms from the periphery of the spot

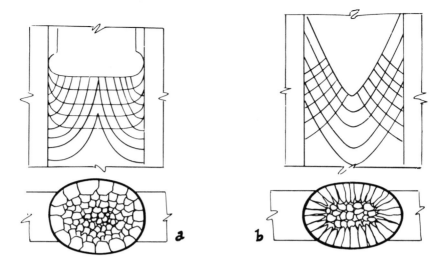

5.13 Structure of vertical electroslag welds, appearing on vertical sections (parallel to surfaces) and transverse sections: a) Flat molten pool: vertical grain growth; b) Hollow molten pool: transverse grain growth.

towards the interior, under the combined effect of intense cooling by conductivity, inside the metal, and towards the electrode tips, themselves extremely conductive of heat and, furthermore, cooled. If, by its own solidification speed the material is suitable, we obtain a radial structure in which the grains develop by joining in the equatorial plane (Fig. 5.14a). In the opposite case, the orientated solidification structure is replaced by an equi-axial structure in which the grains occupy a more or less important proportion of the central part of the spot[12] (Fig. 5.14b).

Role of transformations in the solid state

In a material subject to solid state transformations,[13] which is essentially the case with steel, it is necessary to examine the chronology of the phenomena during the cooling which follows solidification, and the effects of reheating of the weld metal after solidification and cooling, in particular during multipass welding.

As we have said above, metallographic examination of a steel weld reveals, in the weld metal zone, a structure which could be described as solidification-transformation because it results from the superposed effects of the initial solidification (in the form of delta ferrite (δ) or gamma austenite (γ) depending on chemical composition) and from the later solid state transformation, which takes place in accordance with the conditions depending on the law of cooling and the chemical composition. But the constituents arising from this transformation have a morphology inherited from that of the initial solidification structure; also, the shape of the transformation

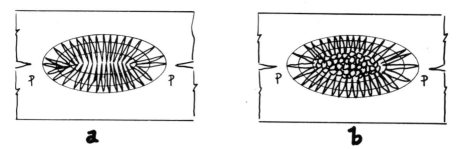

5.14 Solidification structure of resistance spot welds: a) Radial grain growth continuing as far as the diametral plane; b) Interruption of radial growth with appearance of an equiaxial structure. p peripheral zone hot pressure welded.

isotherms, although different, is close enough to that of the solidification isotherms.

For these two reasons, although a detailed examination shows that they do not coincide, the solidification and transformation structures are quite close to each other with regards to their orientation and morphology. In particular, the epitaxy of the fusion boundary occurs as for solidification. We see this, for example, in Fig. 5.15: the intergranular system of the pro-eutectoid ferrite holds during passage

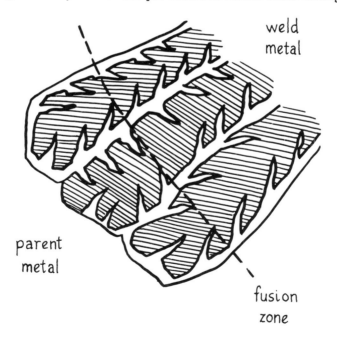

5.15 Passage from solidification structure to transformation structure at fusion boundary (steel). The pro-eutectoid ferrite which precipitates at the grain boundary and inside grains on either side of the fusion boundary maintains the epitaxial appearance.

between the parent and solidified and transformed weld metal, together with the orientation of the intergranular ferrite.

As a practical consequence of this phenomenon, it can be concluded that every factor which refines the solidification structure (chemical composition or heat cycle) also refines the transformation structure. We shall see later that it is also possible to reduce the effects of structural heredity by intervention on the nature of the transformation to obtain finer and more dispersed constituents.

As for transformation on heating, its existence in steel makes it possible, for the weld metal zone of fusion welds, to obtain a grain refining effect comparable with that obtained when annealing moulded parts. This effect occurs not only during post-weld heat treatment (for example to refine the naturally coarse structure of a high energy weld), but also and especially with multipass welds. Indeed, we have seen, during our study of the heat cycle, that each run reheats the surrounding metal, particularly that of the preceding run. Every part of that run in which the temperature of point A_3 is exceeded undergoes the phenomenon of structural regeneration, by the passage to the austenitic state on heating, followed by renewed transformation on cooling, which obliterates and replaces the earlier structure.

Thus, whilst a multipass weld on metal free from transformation or a single phase alloy (Al, Ni or even austenitic steel) comprises only solidification metal unmodified during successive runs (Fig. 5.16a), the same weld, carried out on non-alloy or low alloy steel under the same conditions, includes areas of annealed weld metal separating zones of solidification-transformation metal (Fig. 5.16b). Transposed into the practical case of a multipass weld on thick metal place, this explanation makes it possible to understand, for example, how the weld metal zone of a

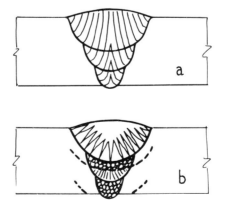

5.16 Role of transformation on heating during multipass welding: a) Non-alloyed metal or monophase alloy: maintenance of solidification structure during successive passes; b) Steel: intervention of transformation on heating: structural regeneration in the portion of each pass where temperature A_3 has been exceeded during the following pass.

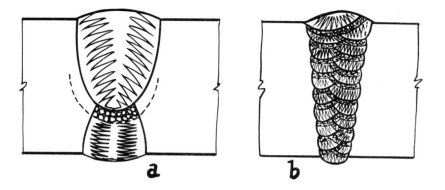

5.17 Application in the case of a multipass weld on thick metal plate.

submerged-arc weld carried out in two runs (face and reverse side) includes a low proportion of metal regenerated in this way (Fig. 5.17a) whilst a MIG or MAG or multipass submerged-arc weld on a narrow chamfer is, on the contrary, almost totally constituted by regenerated metal (Fig. 5.17b). The consequences of this finding will be raised again with regard to the mechanical properties of the weld metal zone.

Physico-chemistry of solidification

Segregation

The phenomenon of segregation, i.e. the local heterogeneity of chemical composition of the product obtained, can accompany solidification at three levels, namely:

- at the crystallisation level: dendritic segregation,
- at the grain level: intergranular segregation,
- at the macroscopic level: major segregation.

The section indicates in what measure the phenomenon, well known elsewhere (in foundry and in iron and steel metallurgy in particular), can affect weld solidification.

In alloys, as in impure products, dendritic segregation results both from the mode of crystal growth, and from the 'constitutional supercooling' phenomenon. The growth mode of crystal structures, already mentioned, is dendritic, i.e. the crystals tend to develop from nuclei, in well defined directions associated with their crystalline structure (perpendicular to the sides of the cube for cubic crystals); the direction which is close to the solidification path is therefore favoured (Fig. 5.18). In a

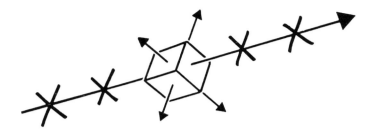

5.18 Diagrammatic representation of the construction of a solidification grain from an initial cubic crystal: growth axes arranged perpendicularly to the cube sides.

pure metal, this mode of solidification is obviously not accompanied by any chemical heterogeneity, each orientated edifice supplying a grain, in which the dendritic crystallisation is difficult to reveal. In an alloy, for example of a solid solution, or an impure metal, the first solidification nuclei to appear produce an evolution in the chemical composition of the liquid containing them, thus retarding solidification. This is the way constitutional supercooling occurs. This phenomenon, renewed step by step, results in the construction and growth of dendrites of variable composition from their central, less easily melted, part towards the outside, more easily melted, part. Because of the speed of cooling, homogenisation by diffusion remains incomplete, and chemical heterogeneity remains after cooling, which makes the dendrites obvious under metallographic examination, and which can be quantitatively complemented by chemical analysis, for example by the Castaing micro-probe.

In fact, in welding, this mechanism does not occur at the start of the solidification path, i.e. at the fusion boundary. In effect, the speed imposed on solidification is momentarily very low and the temperature gradient very high. There appears at that spot a narrow layer of structure called planar (Fig. 5.19), the grains forming along a continuous front, dendrites only appearing further on, when the speed imposed on solidification begins to increase.

Intergranular segregation occurs, for alloying elements as for impurities, in the final stage of the formation of the crystalline structures which constitute the grains. In effect, the mechanism of repulsion towards the exterior of a more fusible liquid continues to the limit of each structure, i.e. to the boundaries of the grains, where pinpoint chemical analysis may reveal local heterogeneity concerning either an alloying element, or especially a fusible constituent resulting from an impurity.[14] This form of segregation also affects the planar structure zone where the grain joints are an extension of those of the substrate, which is particularly significant as regards the risk of hot cracking.

With regard to major segregation, there is a need to examine what form this phenomenon can take in welding. This type of segregation (already mentioned in Chapter 4), concerns, at the time of solidification of an immobile liquid mass[15]

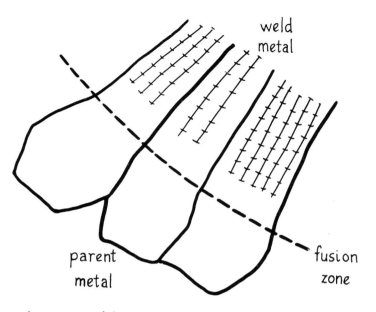

5.19 Micrographic appearance of planar structure on commencement of solidification on the fusion boundary; the dendritic appearance appears only after a certain distance from the solid/liquid interface.

(ingot or moulding) the heterogeneity of chemical composition of the part, in which the central section is richer in elements (additions or impurities) lowering the fusion point, and therefore repelled towards the inside as solidification progresses from the outside.

Thus defined, major segregation can only occur, strictly speaking, in spot welding since it is the only process involving an immobile molten pool. Effectively, the solidification mechanism (already described and illustrated in Fig. 5.14) implies the possibility of a concentration of fusible elements in the central part of the fusion zone, i.e. in the region where the grains orientated according to arrangement a, or in the form of intergranular segregation in the equi-axial structure of arrangement b, unite. For arrangement a, this segregation could be harmful, because of the cavity shrinkage effect (comparable with hot cracking) which it encourages, if pressure is not maintained at the moment of solidification or subsequent forging.

As for fusion welding processes, we need only refer to the arrangements illustrated in Fig. 5.12 to see that the more transverse are the solidification paths, i.e. the more the solidification grains converge on the centre line of the bead (arrangement b), the more possible is heterogeneity of the chemical composition at that spot where, for each position of the molten pool, solidification reaches completion. In the case of arrangement c, segregation takes on an intergranular character in relation to the equi-axial grains. The phenomenon does not occur for arrangement a, more favourable in this respect. This aspect of major segregation,[16] which can be described as progressive, is found also, depending on the process procedure used, in vertical electroslag welds, the arrangement in Fig. 5.13 being the most favourable, because the impurities repelled by solidification rise towards the slag.

Other causes of heterogeneity

When the filler metal is different from parent metal, homogenisation of weld metal resulting from the mixture of the two is achieved according to welding conditions and, in particular, the method of transfer of filler metal when arc welding. So we find, on a longitudinal section of the fusion zone, more or less recurrent variations in composition. In particular, in arc welding of non-alloy or low alloy steel with Cr-Ni stainless steel covered electrode, the drop transfer of filler metal causes the appearance (Fig. 5.20) of successive alternate bands of dilution (on longitudinal or transverse sections) where parent metal eroded by the electric arc (martensitic bands) and the filler metal supplied by successive drops (austenitic bands) predominate, richer in alloying elements.

Another type of heterogeneity which can be observed lies in the fact that along the length of its displacement, the molten pool collects on its edges the various impurities or inclusions which the parent metal may include. If those impurities are fusible and soluble, they will be found in the segregated state in the solidified metal, as is the case for steel scale, which may also cause cracking of the weld metal. If they are insoluble, and especially refractory, they gather in the molten pool then run aground on the edges of the weld bead; they can modify the shape of those edges by their effect on the surface tension of the weld metal or ionisation of the arc. This can be the case with automatic TIG or MIG welding of stainless steels (Fig. 5.21) because of the rare earth oxides involved in the preparation of those steels.

Practical consequences

Evolution of gases: porosities

As has just been described, the process of solidification of molten weld metal determines the evolution of the gases (or vapours) it may contain, evolution manifests itself in porosities if those gases have no opportunity of escaping the molten pool before being imprisoned in the solidified metal. The gases or vapours in question may arise from influences outside the welding operation itself or even from the behaviour of the molten pool. In the first category, we can mention the breakdown or combustion of coatings, paints or various coverings, the reduction of laminated products scale, the effects of humidity or simply that of air imprisoned when badly prepared or badly executed welds are performed (for example, a penetration defect when welding on the reverse side). As examples of the second category, we can mention the effects of volatilisation (or burn up) of elements (such as zinc in brasses), moisture contained or adsorbed by filler products (release of hydrogen) and by the reaction of effervescence (release of carbon monoxide).

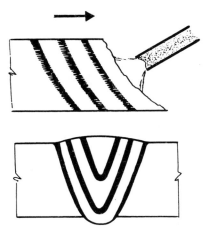

5.20 Macrographic appearance on longitudinal and cross section of periodic heterogeneity resulting from the drop transfer in the case of a heterogenous weld.

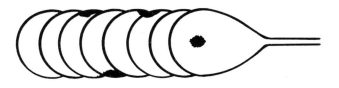

5.21 Formation of refractory particles in the molten pool and separation along the edges of the bead.

Depending on their origin and the conditions of solidification, porosities take various forms, described with the corresponding terminology in French Standard A89-230*.

In fact, the explanation of porosity morphology depending on solidification conditions of the weld metal is especially applicable to the second category porosities mentioned above. This can be done (Fig. 5.22) by using the information given by Chalmers** on the aspects assumed by a release of gases as a solidification front advances pushing the liquid product in front of it. In that situation, Chalmers explains that the conditions of surface tension cause the formation of gas bubbles on the front itself and that those bubbles enlarge and break away by gravity if solidification does not progress too quickly (case a); or indeed they are overtaken by the solidification front if solidification progresses more rapidly (case b); or again, the bubbles, constantly fed by gases, grow by stretching out in the direction of the movement of the solidification front (case c).

---

* NF A89-230: Classification of defects in assemblies welded by fusion with explanatory comments
** Chalmers: Principles of solidification, Publ John Wiley & Sons, 1967

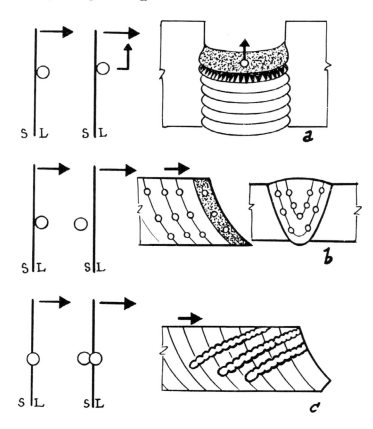

5.22 Chalmers sketches (left) relative to the circumstances of formation and shape of pores and application (right) to pores in the weld zone: a) Slow progression of the solidification front: production of gas bubbles which escape entrapment by solidification; b) Rapid progression of the solidification front compared with the growth of the bubbles, which are trapped by solidification: spheroidal pores, arranged according to the successive positions of the solidification front; c) Continuous supply of gas as the solidification front progresses: vermicular pores developing along the solidification paths.

Transposed to welding, these three situations find their application depending on processes and procedures and also depending on the origin of the gas in question: situation a) is the one which permits any gas bubbles to escape from the molten pool before becoming trapped by the advancing solidification front there are no porosities. The most obvious application, illustrated in Fig. 5.22a, concerns vertical electroslag welding, in which degassing is provided both by the slowness of the progression of the solidification front and by the vertical arrangement which favours gas separation by gravity. It is still necessary for solidification to take place as per diagram a) of Fig. 5.13 and not diagram b).

Situation b) of Fig. 5.22 corresponds to the porosities which the standard (already mentioned) terms 'spheroid' and describes as 'evenly distributed in the weld metal'.

We are talking here about gas bubbles which have been overtaken by the solidification front after forming on it. Effectively, if we examine such porosities on longitudinal or transverse sections, we can see that they are arranged according to the sections of the successive positions of the molten pool. This arrangement has been observed in particular on Al-Mg alloy MIG welds in which porosities arose from moisture adsorbed by the filler metal skin. During welding, a semi-periodic cycle is produced, resulting from a phase of progressively increasing hydrogen content in the molten pool, followed by the formation and entrapment of bubbles, the process recommences.

As for situation c) of Fig. 5.22, this gives rise to 'vermicular' porosities. Their formation implies a continuous supply of gas, either by discontinuity of solubility on solidification (hydrogen), or by a reaction producing a gaseous discharge (carbon monoxide). The corresponding diagram on the figure takes account of the fact that vermicular porosities spread according to the solidification paths. We see this particularly clearly on longitudinal fractures in the weld metal; we see also that vermicular porosities are made up of a succession of joined spheroidal porosities and, in the case of steel, we note that the walls are glossy because of the reducing character of the gases which have caused the phenomenon (hydrogen or carbon monoxide). With radiographic inspection, vermicular porosities are revealed clearly and their correlation with the solidification structure is evident (compare Fig. 5.23 with Fig. 5.12b), the pattern illustrated sometimes being called 'herring bone'.

In conclusion, we can note that the relationship just described between the creation and morphology of porosities, on the one hand, and solidification conditions on the other, shows clearly the important role played by welding procedures.

Hot cracking of the weld metal[17]

All solidification of a metal or alloy is accompanied by shrinkage; the weld metal zone is no exception and immediately behind the molten pool the metal which has just solidified is subject first to this solidification shrinkage, then to the purely thermal shrinkage superimposed. Compensation for this shrinkage by movement of the parent metal is never complete (restraint), so that high temperature plastic deformation is imposed gradually on the weld metal, both

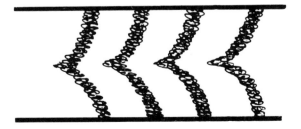

5.23 Radiographic appearance of vermicular pores. Arrangement called herring bone.

117

transversely and longitudinally. In the light of what has just been written on solidification structure, and chemical heterogeneity which can accompany solidification, we see that it is possible for a weld bead to contain two sites which constitute zones of weakness where hot cracking can occur because of the inability of the metal to withstand the deformation imposed on it. These are the central section of the bead (longitudinal cracking) and the boundaries of the grains in the dendritic structure (transverse intergranular cracking). In both cases, when cracking occurs (Fig. 5.24), it is associated with the appearance taken on by major segregation and intergranular segregation in weld beads.

Longitudinal cracking, easily distinguishable on transverse sections, always detectable by ultrasonic inspection, less easily detected by radiography, may or may not extend to the surface of weld beads, depending on whether this cracking is more or less like a shrinkage cavity. For the same material it is especially associated with configuration b) of Fig. 5.12 for here segregation is maximal, whilst it is low or nonexistent for configuration a). In other words, it more readily affects fast welds with a lively molten pool. However, although it corresponds to a high speed configuration c) it is less unfavourable because of the equi-axial structure which, in the centre of the fusion zone, replaces the discontinuity of crystal orientation which characterises the arrangement. If there is cracking, it is of intergranular character.

Transverse intergranular cracking is also encouraged by a welding condition favouring arrangement b) of Fig. 5.12. We have already indicated that due to epitaxy in the fusion boundary, intergranular cracks often extend cracks in the parent metal caused by liquation, i.e. separation, to the grain boundaries, of a liquid phase detracting from hot ductility.

Finally, whether it is a question of porosities or hot cracks, we see that they are more readily associated with certain types of solidification structure than others, i.e. with the procedures used. Hence the importance of choosing suitable welding conditions together with the use of suitable materials and filler products.

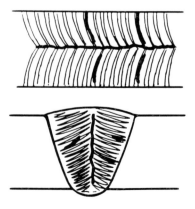

5.24 Macrographic appearance (transverse section and plan view) of transverse interdendritic and longitudinal hot cracking.

Mechanical properties of weld metal

Due to the particular conditions of solidification and the resulting structures, it can be expected that the weld metal zone of fusion welds presents particular features, from the point of view of its mechanical properties, vis-a-vis the parent metal. To take account of these particular features, it is necessary to consider separately the materials susceptibility to transformation points and the case of single or multipass welds.

If, because of its chemical composition, the weld metal has no transformation point, we can in an initial approximation[18] consider single and multipass welds at the same time since multipass welds (Fig. 5.8) are made up by structures of grains orientated one on another by epitaxy. This grain orientation, common to successive runs, creates directionality of mechanical properties which results in different grain performance depending on whether they are stressed in their direction of growth or in the perpendicular direction. We can see this directionality on the section of a lengthwise cut cylindrical tensile test piece (in relation to the direction of welding) in the weld metal, which becomes oval in shape throughout the plasticised zone and especially in the necking down area (Fig. 5.25) whilst the surface of the test piece becomes rough due to the emergence of the sliding planes of the solidification orientated grains. This is the sign of higher ductility in the grain growth direction than in the transverse direction. The same happens for the weld metal of vertical electroslag welds, whose notch impact value depends on the situation of the notch vis-a-vis the grain growth direction (Fig. 5.26). The value is higher for specimens taken longitudinally in this direction.

This anisotropy has no harmful consequences[19] vis-a-vis mechanical properties, at least as far as the majority of metals and their solid solutions are concerned where the crystal structure is of the cubic type with centred sides, which gives them a high degree of ductility[20] even at low temperature because of the slippage possibilities offered by this structure. On the other hand, for the centred cubic sys-

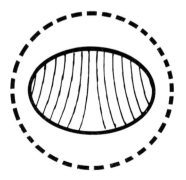

5.25 Consequence of anisotropy of the weld metal of a single or multipass weld on a product free from transformation point (see Fig. 5.8). A tensile test specimen taken longitudinally from the weld metal deforms more, during the test, in the lengthwise growth direction of the grains than in the transverse direction: the test specimen becomes oval.

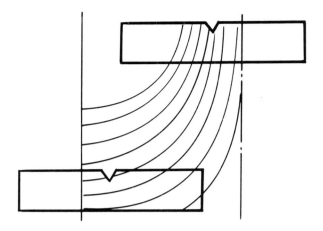

5.26 Possible positions of an impact toughness test specimen and its notch in relation to the growth direction of the solidification grains (example of electroslag welding).

tem, therefore for iron, it is breaking by cleavage which predominates at low temperature, and therefore poor notch toughness if the weld metal has a coarse, orientated structure. This is unfortunately the case for ferritic chromium steel[21] (with 17% Cr and over, depending on carbon content) and the reason why they are fusion welded with austenitic filler metal. We shall see below that the susceptibility of ferrite to cleavage also dominates the problem of mechanical performance of the weld metal in non-alloy or low alloy steel welds. In the case of non- or low alloy steel, a single run weld is characterised after solidification and cooling by a structure resulting from the newly solidified metal transformation in the solid state. As we have said above, this structure presents, to a large degree, characteristics resulting hereditarily from the solidification structure, particularly with regard to constituent size, including ferrite. The result, from the point of view of notch toughness, is performance which, with identical chemical composition, is poorer the bigger (and therefore slower cooling) the molten pool is. This is the case with high energy submerged-arc welds and vertical electroslag welds. To improve notch toughness, with identical heat conditions, it is necessary to concentrate on chemical composition to refine the solidification structure and also the transformation structure (Ti, Mo or V additions) to change their nature by avoiding ferrite blocks.

As for multipass welds, it is obvious that the structural regeneration caused by successive heat cycles is beneficial from the notch toughness point of view. To prove this, we can compare the two situations illustrated in Fig. 5.17: if a high notch impact value is required in the weld metal, it can be achieved straight away for arrangement b). Situation a) requires normalising heat treatment after welding.

With covered electrode arc welding, or MIG or even submerged-arc welding, the notch toughness specifications concern multipass welds. That is why the corresponding standards specify very precise procedures, including in particular temperatures between runs, so as to guarantee reproducible test results.

Chapter 5 **Comments**

1 Sometimes called the 'nugget' in resistance welding.

2 In the sense that even if it is agitated, it does not move.

3 On the reverse side, the molten pool is in contact with solid metal (parent metal, backing plate or metal already deposited) with a gas (including air) or with a solid flux which maintains it.

4 It is easy to visualise this behaviour as locally contaminating the molten pool by a momentary input of extraneous metal in front of the molten pool; for example, an experiment can be carried out by a short input of copper to an aluminium pool.

5 This is why it is incorrect to call it an 'ingot' as is sometimes done.

6 This arrangement, illustrated by Fig. 5.7, which is conspicuous in pure metals or single phase alloys devoid of transformation points, is not always so obvious, either because of cooling transformations, or because of chemical heterogeneities which may exist in the fusion boundary.

7 This behaviour is itself associated with dendritic solidification, to which we shall return later in this book.

8 For some alloys, this mechanism is replaced by the one involving *twinning*, namely a change of crystal orientation in one part of each grain during growth, which enables it to adapt to the orientation imposed by the form of the solidification path.

9 The configurations which follow appear much more clearly on automatic welds than on manual welds due to the irregularities of the operator's movements. Also, for pure metal and single phase alloys, these configurations are indirectly revealed by the diffraction marks visible on radiographs. The corresponding images should not be confused with defects (see Chapter 10).

10 Contrary to what is often described, this is not an equi-axial structure (i.e. with no privileged development orientation). The difference between the periphery of the fusion zone and the central part is simply due to the position of the section in relation to the solidification paths.

11 On account of the influence of this structural difference on performance during welding (risk of cracking), the practices recommended for vertical electroslag steel welding take account of a value of the width/depth ratio of the molten pool surface.

12 The solidification structure of the spot weld is also revealed by the appearance of the fracture during the torsion test, which also reveals a peripheral zone p (Fig. 5.14) where a pressure weld has preceded fusion in the central part. Hence the interest of this test in determining welding procedures (see Chapter 10).

13 Solid state transformations are examined in more detail in Chapters 6 and 7.

14 Thus it is, in the first case, chromium sulphide in Cr-Ni stainless steels, or nickel sulphide in welding of this metal.

15 In the sense already explained, i.e. prepared in a single operation and not step by step.

16 A mechanism closely related to major segregation (and to shrinkage cavities) occurs systematically in the terminal craters of weld beads.

17 As we shall see later, in steel welding weld metal is not exempt from cold cracking. But the phenomenon is not in direct relation with the solidification structure.

18 Subject to possible precipitation under the influence of the heat cycles of successive weld runs (for example, chromium carbides in austenitic stainless steel welds).

19 It may, on the other hand, have harmful consequences from the point of view of corrosion.

20 Ductility which does not however reach that of the equi-axial structure of the parent metal, therefore is isotropic.

21 However, a coarse orientated structure does not of itself generate brittleness, in the absence of impurities precipitated in the crystallographic planes concerned. This is how we know how to produce ferritic steels with low carbon and impurity contents, much more weldable than normal ferritic steels.

6 Solid phase transformations during welding (heating)

General points

This chapter is devoted to the solid phase transformations (in the most general sense of the word 'transformation') which occur in the parent metal, and also, possibly in the weld metal (in the case of multipass welding) during the heat which accompanies welding. These transformations concern constitution and structure in the sense that we have already given to these terms and which we need to go into in more detail.

For a metal or alloy, the term 'constitution' is reserved for the nature and proportion of the phases whose presence is revealed and identification made possible by the investigative facilities available to the metallographist, on a microscopic or crystal scale. So, for example, steel in the annealed condition is made up when cold of ferrite α and cementite Fe_3C, present in the proportions indicated in the iron-carbon equilibrium diagram for the carbon content of the steel under examination. Thus defined, constitution is dependent on temperature and this is precisely the object of the equilibrium diagrams to take account of this influence. Changes in constitution are reversible if the temperature variations are compatible with the state of equilibrium. If such is not the case, returning to ambient temperature results in the appearance, in an unbalanced state, of constituents different from the initial constituents or, on the borderline, by the maintenance of the constitution predominating at the temperature reached during heating. So, in the case of steel, the constituents α and Fe_3C existing at ambient temperature give way to the austenite solid solution γ from temperature A_3 (of the order of 900 °C depending on carbon content). If subsequent cooling is sufficiently slow, the ferrite and cementite will be found again in the initial proportions at ambient temperature; if not, the equilibrium conditions will not be respected and the final constitution will be different on account of the hardening phenomenon which can lead to martensite, an unbalanced constituent.

When we talk about structure and more precisely micrographic structure,[1] we mean the distribution and morphology of the constituents present. A grain is coarse or fine, equi-axial or orientated, or again, a precipitate is bulky or finely dispersed. Thus defined, micrographic structure is influenced by temperature (and also by time), but in an irreversible manner. For example, in annealed steel, we find ferrite and cementite (constituents) in the aggregate called pearlite, where both constituents are arranged in alternating lamellae. Prolonged heat treatment at a temperature close to but lower than temperature A_1 causes coalescence i.e. the cementite lamellae become globular. Contrary to what takes place for constitution, this variation of structure is irreversible in the sense that after cooling cementite is found in the form which has resulted from soaking at the temperature of the heat treatment. Also, in addition to its sensitivity to heat treatments, micrograph structure is sensitive to mechanical treatments, hot as well as cold.

This distinction between constitution and structure does not imply that these two characters are independent. Structure, in particular, influences the conditions in which constitution is established, during heating as well as cooling. This is particularly the case for the phenomenon of overheating of steel (in the austenitic state); this phenomenon acts on the cooling transformation mechanism, and thus has an effect on structure and constitution after cooling.

Since constitution and structure are affected by temperature variations, we must expect to see the effects in and around welds. We must not, however, in interpreting these effects, lose sight of the brevity of most welding cycles. This chapter is devoted to heating transformations, both for structure and constitution. Chapter 7 deals with transformations on cooling.

Modifications of structure

Recrystallisation after work hardening

The term recrystallisation means reorganisation of the crystal lattice and consequently the grains of a metal (or alloy), by means of heating from a work hardened state. We say that a metal is work hardened when it has undergone cold working which has resulted in a modification of its micrographic structure (the grains have deformed in the direction imposed by the mechanical treatment) and of its crystal structure (sliding against each other and distortion of planes of greater atomic density). These modifications are accompanied by an increase in resistance to deformation (tensile strength, yield strength, hardness) and a reduction in deformation capacity (elongation, necking down, impact toughness). As shown in Fig. 6.1, this variation depends on deformation rate e%, or degree of work hardening which is limited to the value for which deformation capacity becomes nil.

If a work hardened metal is subjected to heating, we see the recrystallisation phenomenon, the successive phases of which are illustrated in Fig. 6.2, on a micrographic structure scale, in relation to grain size and evolution of mechanical properties.

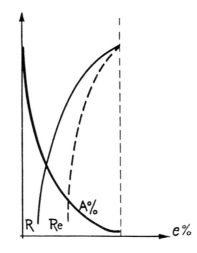

6.1 Diagrammatic representation of the variation of tensile properties generated by work hardening (R impact strength, R_e yield strength, A% elongation).

From the work hardened structure (diagram 1), we see, close to the recrystallisation temperature θ_r the appearance of nuclei[2] (diagram 2) developing in an equi-axial structure (diagram 3), in which the higher the degree of initial work hardening, the finer grains g are. This structure, called primary recrystallisation, stabilises with a slight enlargement of grains up to temperature θ_s, where overheating commences (also called secondary recrystallisation).

At the same time that recrystallisation is in progress, mechanical properties are evolving in the opposite direction to the effects of work hardening, i.e. the metal regains its elongation capacity, whilst the strength given to it by the work hardening diminishes.

The phenomenon of post-work hardening recrystallisation has a direct application in welding on work hardened products. In this case, every welding operation gives rise to softening of the parent metal as far as the recrystallisation temperature isotherm is reached. This effect, apparent on hardness relationship (Fig. 6.3) appears more or less vis-a-vis tensile characteristics recorded on a welded test piece machine perpendicularly to the joint, depending on the width of the zone affected. In the case of a relatively wide zone (for example TIG or plasma welding), reduction in hardness corresponds to a reduction in tensile strength, breakage effectively occurring in the softened zone because deformation can develop there. If this zone is narrow (for example in EB welding), the break is only localised there with a higher load because the neighbouring hard metal constrains the softened zone; it can even not be localised there at all, and the effect of the softening does not show.[3] The fusion zone also has a role to play through its possible bracing/stiffening effect, depending on whether it is itself soft like the recrystallised metal or hard like the work hardened parent metal.

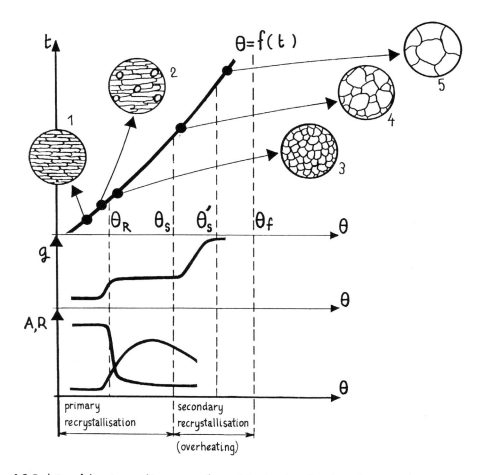

6.2 Evolution of the micrographic structure of a work hardened product during heating and corresponding variations of grain size and tensile characteristics.

Thus occurs the phenomenon of recrystallisation generated by welding in the case of products free from allotropic transformation, such as aluminium or austenitic stainless steel. It is a different matter for non- or low alloy steel because of the involvement of transformation α/γ which occurs at the temperature of point A_3, i.e. a little above the ferrite recrystallisation temperature $\theta_{R\alpha}$. In this case, (Fig. 6.4) the zone softened by the ferrite recrystallisation is limited between $\theta_{R\alpha}$ and A_3 because, from this latter temperature, reaction $\alpha \rightarrow \gamma$ produces, by another mechanism, the effect of normalisation which limits the width of the softened zone. In this way, there is generally practically no fear of recrystallisation softening when welding a work hardened steel other than extra-soft. In any case, with welding in the work hardened condition, it is always possible, if the thickness of the welded metal permits, to reharden by cold hammering the zone softened by the welding cycle, whether or not the alloy involved is transformable.

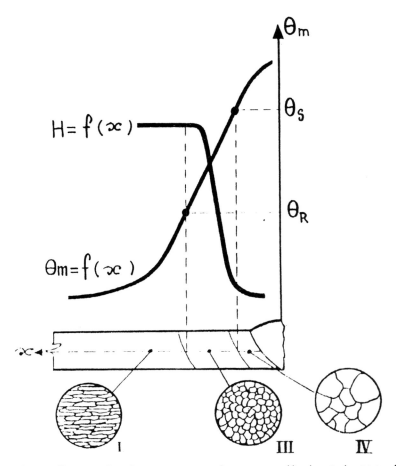

6.3 Impact of recrystallisation and overheating on micrographic structure and hardness in the vicinity of a weld on a work hardened product. The Roman figures refer to the structures indicated in Fig. 6.2.

So we can use to advantage the phenomenon of post-work hardening recrystallisation in various circumstances which are useful to know, in particular for acting on the arrangement of inclusions and also for modifying multipass weld structure where the weld metal has no allotropic transformation point. Indeed, for the first case, the interest of pre-weld hammering lies in the fact that recrystallisation does not affect inclusions. Consequently, intergranular constituents or inclusions, liable by their melting temperature to create liquation problems around welds (appearance of a liquid phase at grain boundaries), are separated from grain boundaries when the recrystallisation caused by welding occurs, because of prior work hardening. Figure 6.5 shows the chronology of operations and phenomena generated: the edge to be welded is first hammered (I), which replaces the equi-axial structure with precipitates at the grain boundaries by a work-hardened structure, where the precipi-

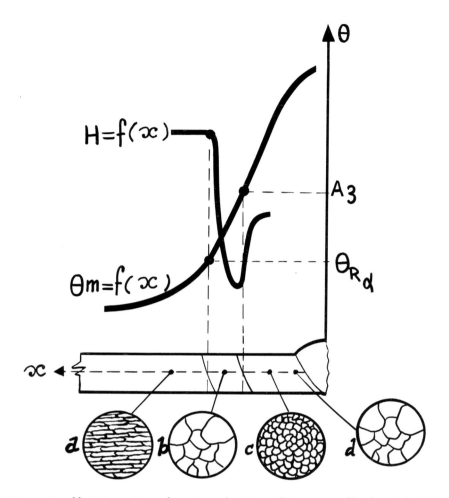

6.4 Intervention of ferrite/austenite transformation on the micrographic structure and hardness in the vicinity of a weld on work hardened mild steel: a) Work hardened ferrite; b) Recrystallised ferrite; c) Ferrite resulting from austenitisation a little above A_3; d) Ferrite resulting from overheating of the austenite.

tates are more or less broken and elongated by deformation. The subsequent welding cycle (II) generates a recrystallisation grain in the hammered part, the inclusion remaining in place, i.e. independent of the new grain boundaries. Better still, this result can be obtained by prior 'buttering' (III) i.e. surfacing of the edge to be welded to produce recrystallisation prior to the actual welding.[4]

A second case of practical utilisation of recrystallisation after work hardening is that of cold hammering between passes (Fig. 6.6) where multipass welds are being made, when the weld metal has no transformation point (for example Al, Cu or even austenitic stainless steel). In such a case, if there is no intervention between passes, epitaxy produces the crystal structure already described (I); on the other

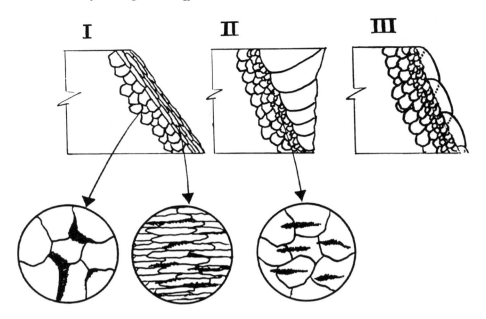

6.5 Utilisation of recrystallisation by prior hammering of the contact edges: I – preparatory hammering phase of edges to be welded; II – recrystallisation caused by welding, separation of grain joints and inclusions; III – intervention of buttering ensuring recrystallisation before the actual welding.

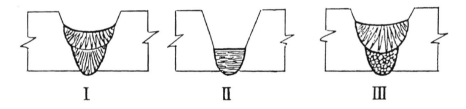

6.6 Recrystallisation of the weld metal by hammering between passes in a metal free from transformation points: I – two successive passes without intermediate hammering; II – hammering between the two passes; III – the second pass causes recrystallisation of the first.

hand, if hammering is done to work harden the metal which has just been deposited (II), the heat cycle of the following pass produces by recrystallisation an effect (III) quite similar to that of the normalisation by successive passes in non-alloy steel welding. So, as in the case of steels, we can thus obtain an equi-axial structure over a good proportion of the joint, with a corresponding improvement of mechanical properties, at the same time as mechanical stress relief with each hammering. The grain refining effect in the fusion zone can also be achieved for single pass welds, but in two successive operations, one of localised hammering of the fusion zone and its edges, the other of heating the zone thus hammered by means of an exterior heat source.

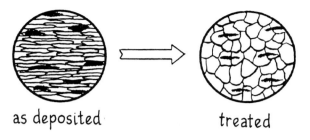

as deposited   treated

6.7 Recrystallisation of a metallic layer deposited by projection, after heat treatment.

In the same way, heat treating a component covered with a thermal sprayed metal coating (metallising) produces recrystallisation of the deposit which reconstitutes in equi-axial grains, independently of laminated oxide inclusions which have been able to form during spraying (Fig. 6.7).

Simultaneous deformation and recrystallisation

Recrystallisation temperature, which is dependent on the degree of workhardening and the purity of the metal[5] separates the area of cold deformation, below that temperature, from that of hot deformation, which is above that temperature. In this temperature range, recrystallisation nuclei appear at the same time as deformation progresses and they constantly produce a recrystallised structure and regenerated grains. The same is true in hot rolling or forging.

This characteristic is used to advantage in achieving the metal continuity involved in welding processes in the solid phase with pressure and heating (hot pressure welding), such as gas pressure welding, by induction, resistance, friction or even by diffusion. In these processes, the temperature reached is sufficient for recrystallisation from common nuclei on both contact surfaces to establish metal continuity. In diffusion welding, the phenomenon occurs only partially, from a few rough spots on the contact parts, but it is enough to establish 'bridges' by which the diffusion which completes the bonding process is initiated.

As for the cold welding processes which involve the deformation required to bring the atoms together, these obviously generate workhardening of the interface, but also, to a certain extent, localised recrystallisation of the contacts, by means of the heating produced by the speed of deformation. Hence the, at least partial, occurrence of recrystallisation in the bonding process. Subsequent heat treatment completes the recrystallisation.

Another consequence of spontaneous recrystallisation during hot deformation appears, during fusion welding, in the nearby parent metal of the fusion zone. Indeed, the temperature gradient existing in that zone gives rise to a more or less intense plasticisation, at a temperature permitting simultaneous recrystallisation.

Thus, even in the absence of transformation during heating, a change in grain form and size may occur irreversibly due to the effect of welding, especially if the recrystallisation temperature is low (pure metals).

Overheat

The last phenomenon before fusion is what we commonly call 'overheat' but metallurgists describe as 'secondary recrystallisation' in so far as, in the scale of temperature, it follows immediately upon recrystallisation and the moderated grain enlargement which accompanies it. Indeed we see (Fig. 6.2) from the temperature θ_s that certain grains disappear to the benefit of others which grow so that above a second temperature we find only large grains. This phenomenon, associated with grain boundary energy and movement of those boundaries, is quite different from primary recrystallisation, in the sense that it corresponds not to the germination of new grains but to an increase in size of existing grains, without any change in their orientation. The dimension of the overheat grain depends on temperature and time at that temperature[6] and also on the degree of initial work hardening. Further more, some inclusions are liable to limit overheating by curbing grain boundary movement. Grain and overheat temperature depend on the degree of initial work hardening, but each in a different sense as shown by the representation (Fig. 6.8) proposed by Portevin*, where the effects of the two phenomena,

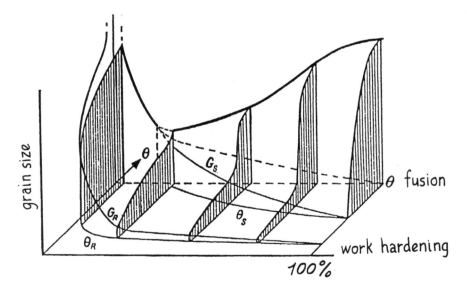

6.8 Evolution of grain size (recrystallisation then overheating) during reneating after work hardening: θ_R recrystallisation temperature, G_R recrystallisation grain, θ_s start of overheat temperature, G_s start of overheated grain.

---

* PORTEVIN: Introduction to the study of heat treatment of metallurgical products. Special steel, metals and alloys –Paris, 1934

recrystallisation and overheating, are superimposed, as regards temperature and grain size. We see that, with low degrees of work hardening, recrystallisation and overheating merge.

This is indeed what we see around welds, whether the parent metal is work hardened or hot deformed by the welding operation. To varying degrees depending on the metal or alloy in question, the grain continues to grow when the distance in relation to the fusion zone decreases (Fig. 6.3), at least in the absence of a transformation point. In steel, we have seen earlier that grain evolution is interrupted by the transformation α → γ; so it is austenitic grain which undergoes overheating until fusion.[7] This is the reason why we find in Fig. 6.4 a fine grain structure framed by two coarser structures, the one resulting from recrystallisation and commencement of overheating of the ferrite, the other from overheating of the austenite, which hereditarily generates a coarse ferrite grain on cooling. As for the extent of the overheated zone and the size of the grains, we have during our study of the heat cycle already indicated that they depend on temperature gradient and soaking time at high temperature which characterise each process and, for a single process, the energy involved, hence the overheat structures that we meet in gas welds, submerged-arc welding, especially at high energy, or again in vertical electroslag welds where the phenomenon is encouraged by the relative slowness of the process. Overheating is, on the other hand, limited to arc welding with covered electrodes or gas fluxes, and especially in spot welding. Finally, let us not forget that, amongst the procedure factors which have an influence on temperature gradient and the time at high temperature, there is preheating: preheat temperature should not be raised unnecessarily beyond the necessary minimum because there is no advantage to be gained as regards the object of that precaution[8] and the disadvantage of overheating is accentuated.

As regards the consequences of overheating in welding, they are metallurgical and mechanical. From the metallurgical point of view, an initial consequence is, with the help of epitaxy during solidification, the influence (already mentioned) of the size of overheated grains, through the fusion boundary, on the size of the solidification grains in the fusion zone. Also, any step taken to limit the size of the overheated grain is at the same time advantageous vis-a-vis the solidification grain. Again, the phenomena likely to occur during cooling and which results from diffusion towards the grain boundaries are curbed or cancelled out by overheating. So it is, in steel, with the separation of the pro-eutectoid ferrite which has a tendency to take place inside the grains (Widmanstaetten's structure) with structural heredity, the superheat grain giving a coarse structure after transformation on cooling.

From the mechanical point of view, overheating, although it slightly alters them for the worse, has no important effects on the mechanical properties of the metals with face centred cubic crystalline lattice,[9] such as aluminium, copper, nickel, austenitic steels, etc. On the other hand, for steel, the embrittlement influence is obvious, both because of the body cubic lattice of the ferrite, and because the less abundant grain boundaries play a lesser role in obstructing propagation of fractures. Nevertheless, in normal conditions of use, the minimum characteristics required for the parent metal are normally obtained in the overheated zone of welds. They are not inferior in every case to those of the weld metal in the solidification and transformation state. The same is not so for chrome ferritic steels, which for this reason, are mainly used for relatively thin components fusion welded with Cr-Ni austenitic filler metal.

Coalescence

The term coalescence has already been used in speaking about the phenomenon in which the particles of a constituent present in a solid solution (normal constituent or impurity) are liable, under the influence of interfacial forces, to gather together and become globular during heat treatment. It is often a quite slow evolution (for example in the case of globularisation of lamellar pearlite in steels); but when the particles are finely dispersed, even a brief heat cycle like that of welding is liable to cause coalescence, most often as the final stage of a tempering effect (see following paragraph). Thus it is for stainless steels, some light hardening alloys, maraging steels, etc. It is also the case for micro-alloy steels in which the carbonitride inclusions (in particular titanium) limit grain enlargement, provided they are adequately dispersed and not coalescent.

Modifications of constitution

Discontinuance of out-of-balance conditions, return to solution

Nature of the phenomenon

In the scale of increasing temperatures, the first consequence of heating, vis-a-vis the constitution of an alloy, is the return, from an unbalanced state to a state closer to or even at equilibrium. In a binary or pseudo-binary alloy, such an evolution may occur when the equilibrium diagram shows a variation in solubility, depending on temperature, of a constituent B in a solid solution A, this variation appearing as an oblique line on the equilibrium diagram (Fig. 6.9). Under these conditions, for an alloy of C% composition, heating above θ_s permits constituent B to become a solution in phase A. Conversely, slow cooling permits precipitation of B, whilst rapid cooling maintains B in saturated solution in A, out of balance. From this state which we call hardened, further heating tends to cause phase B to re-appear depending on the conditions that the welding heat cycle is likely to produce. A few examples illustrate this behaviour.

Cr-Ni austenitic stainless steel

For these steels, the diagram in Fig. 6.9 is the one for solubility of carbon in Cr-Ni austenite, the phase playing role B being the $Cr_{23}C_6$ chromium carbide. For the steel in question to be austenitic at ambient temperature, the carbon content must be very low (C $\leqslant$ 0.03%) or the steel must be in the 'hyperhardened' state.[10] But this state is unstable and the welding heat cycle in a

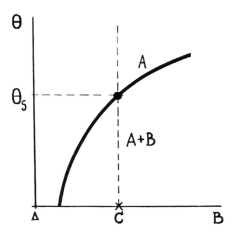

6.9 Arrangement of a pseudo-binary equilibrium diagram making possible a two-phase condition (A + B) by slow cooling, hardened (supersaturated A) by rapid cooling or tempering (A + B finely dispersed) by reheating from hardened condition.

temperature range around 650 °C causes intergranular precipitation of chromium carbide (Fig. 6.10); this makes the parent metal liable to intergranular corrosion in the region affected. Outside that region, the welding cycle first of all causes coalescence of the precipitate, which renders it harmless because of the diffusion, then returns it to solution, followed by renewed hyperhardening if the post-weld cooling is fast enough to cause it. The same applies to the weld metal if it is obtained in a single pass; but the precipitation of carbides is liable to occur if there are several passes, or in the case of crossing of weld beads (see Chapter 7).

It is because of this phenomenon that Cr-Ni austenitic steels destined for fusion welding are either low carbon or stabilised with titanium or niobium, both of these elements fixing the carbon in the carbide state; this avoids precipitation of chromium carbide. The same goes for filler products, with the restriction that the volatility of titanium prevents its use as a stabilising element.[11]

Note also that the requirements relative to spot welding are less rigorous because of the brevity of the heat cycle and intense temperature gradient which are characteristic of that process.

Hardening aluminium alloys

The example of two light hardening alloys, one copper based and the other zinc based, illustrates the different behaviour on cooling, from identical behaviour on heating.

Copper-based hardening aluminium alloy,[12] generally called 'duraluminium' is in the hardened and aged condition. It has in effect undergone a dissolving treatment at a temperature higher than θ_s (several components such as Al_2 Cu, Mg_2 Si, Mg_2 Al_3 play the role of constituent B) followed by rapid cooling which maintains solution A

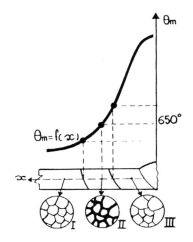

6.10 Precipitation of chromium carbide in the vicinity of a fusion weld on Cr-Ni austenitic steel: I – parent metal initially hyperhardened: solid solution γ; II – sensitised parent metal: solution γ + intergranular carbide; III – parent metal hyperhardened by the welding heat cycle: solution γ.

in the supersaturated state. Renewed heating, at moderated temperature θ_v then encourages ageing, which is an ultra-microscopic precipitation of constituent B; this gives the alloy its recognised high tensile characteristics. Under the effect of the welding heat cycle (Fig. 6.11) the zone where temperature θ_v is reached and exceeded is the seat of coalescence of the components mentioned above, then, from θ_s, the return of those components into solution. On cooling, those constituents are precipitated once again, because hardening does not occur spontaneously; hence the significant weakening of mechanical characteristics, accounted for by the hardness

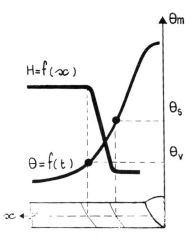

6.11 Hardness relationship in the vicinity of a weld on copper-aluminium alloy: softening from temperature θ_v by coalescence then θ_s by solution annealing.

relationship both in the parent metal and in the weld metal, firstly by coalescence then by return to the equilibrium state. This is the reason why light hardening alloys of this type are considered unsuitable for use in welded constructions.[13]

The case of Al-Zn alloy (for example at 5%) is a little different: softening by returning to solution on heating also takes place, not because of cooling precipitation but as a consequence of hardening (maintenance of B in solution), which is spontaneous. But this softening is temporary because the hardening precipitation also occurs spontaneously, at ambient temperature, depending on time. In this way, the depression of hardening illustrated in Fig. 6.11 gradually disappears when cold, after welding. Unfortunately, as we have seen earlier, the volatility of zinc poses a problem for the fusion zone, either in MIG welding or in EB welding.

Ageing of mild steel

This example can be considered anecdotal since it concerns the influence of nitrogen in steel, whilst modern methods of preparation now supply products with much lower nitrogen content than in the past. However, the phenomenon still exists, even if it is only apparent in a temperature range which renders it less harmful, and its involvement in welding is worth mentioning especially as the weld metal zone is not necessarily exempt.

The ageing of steel is associated with the precipitation of Fe_4N iron nitride (which plays the role of constituent B) in the ferrite. In the annealed state, the nitride remains in supersaturated solution, but heating to a moderate temperature (250 °C) is liable to cause ultra-fine precipitation, especially if cold deformation has occurred previously, or even occurs at the same time as the heating. This precipitation causes embrittlement,[14] but renewed heating at a slightly higher temperature (650 °C) eliminates the effects by means of coalescence of the iron nitride. Transposed to the case of welding, this information shows that there is a possibility of ageing of mild steel every time welding is performed on or near zones which have undergone cold deformation, during manufacture (cold rolled sections) or due to the conditions of use of the products (ferruling, bending, folding, etc). Any zone which has been thus deformed is liable to ageing, in the sense indicated above, when heated by the welding heat cycle to a few hundred degrees.[15] That is why it has long been recommended that welds should not be located near to cold deformed zones in the parent metal.

In any case, it has to be accepted that the welding operation itself causes ageing, in the widest sense of the term, since it is accompanied by simultaneous heating and deformation. This is why a series of notch impact tests along a line perpendicular to a butt weld reveals in the area concerned a maximum transition temperature, which results in relative embrittlement. This tenso-thermal ageing may also occur in the weld metal of multipass welds on thick steel sheet or plate; we then find, at the base of V welds or in the centre of double V welds, higher transition temperatures than those recorded nearer the surface, due to the hot deformation imposed in these zones by the shrinkage of higher passes.

Stress relieving treatment, which takes place precisely in the temperature range where nitride coalescence occurs, reduces or eliminates embrittlement due to ageing. So normalising treatment (i.e. above A_3) ensures more complete regeneration. For example, post-weld annealing of liquid gas containers is as effective in treating welds as for eliminating the risk of ageing after cold swaging of constituent elements.

Tempering and super-tempering of hardened steel

With hardened, for example martensitic, steel we have another type of unbalanced state where the carbon is held in a supersaturated solution in a martensite constituent, which is itself directly derived from the austenite by cooling (see Chapter 7). Starting from the martensitic state, reheating to a temperature lower than that of point A_1 causes the carbon to be taken out of solution and precipitated in the form of very fine Fe_3C cementite particles, giving rise to a 'sorbite' structure. This phenomenon, called tempering, which occurs from temperature θ_R (Fig. 6.12) is accompanied by softening and an improvement in impact value; from temperature θ_c there is coalescence of the cementite, just prior to the return to solution of this constituent, at point A_1. Also, if a steel initially hardened and tempered at temperature θ_R (having therefore reached hardness H_R and impact value K_R) is heated again, the cementite precipitation mechanism recommences from θ_R, with the corresponding evolution of hardness and impact value. This evolution results in what we call super-tempering which occurs in hardened and tempered steel; as for hardening, this occurs during multipass welding of hardening steel,[16] or in exceptional cases, on hardened steel.

When welding steel hardened and tempered at temperature θ_R (Fig. 6.13), the effect of super-tempering appears from the isotherm corresponding to that temperature and continues as far as isotherm A_1, i.e. to the beginning of the austenitised zone, hence a softening which is more or less extensive depending on the temperature gra-

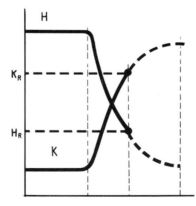

6.12 Evolution of hardness H and impact toughness K during reheating treatment: a) After hardening: tempering effect (solid line); b) After tempering at θ_R: effect of super-tempering (broken line).

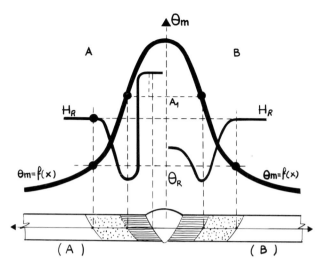

6.13 Hardness relationships in vicinity of a weld on steel hardened and tempered at θ_R: A example of weld hardening steel; B example of steel not hardening during welding.

dient corresponding to the welding conditions used (essentially initial temperature and energy). From the mechanical point of view, the effect of this softening is minimised or accentuated depending on whether the austenitised zone hardens on cooling (case A) or does not harden again (case B). In the first case, the softened zone, which is relatively narrow, is framed and stiffened by two harder zones (the parent metal and the austenitised and rehardened zone); strength is affected little or not at all. In the second case, the softening effects compared with the parent metal (due to the super-temper and absence of hardening) are cumulative over a wide enough zone for strength to be affected. As a practical conclusion to this explanation, it is recommended that, for multipass welding on hardened and tempered steel, welding energy and any preheating be limited as far as possible, especially with a steel which does not spontaneously reharden for the welding procedure employed.

As for the temper itself, apart from the exceptional case of welding on hardened but untempered items, it occurs each time that a pass is accompanied by hardening and a subsequent pass causes reheating which itself produces tempering, in the weld metal as in the parent metal. This is how it is, for example, in the case of austenitic surfacing with wide layers on thick plates of hardening steel (Fig. 6.14).[17]

Allotropic transformation: steel

In the case of steel, the existence of two allotropic varieties of iron, one of which, called austenite γ (face centred cubic lattice) follows, ferrite α (body centred cubic lattice) in the scale of increasing temperatures, stable at ambient temperature, gives rise to a variation in constitution during heating because carbon has very low solubility in ferrite whilst being extremely soluble in

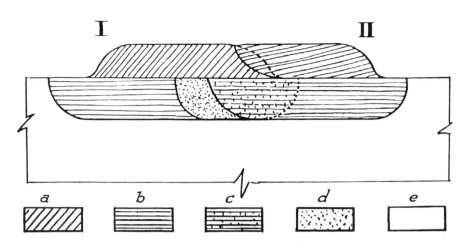

6.14 Macrographic appearance (transverse section at right angles to toe between two wide austenitic surfacing passes on thick hardening steel plate:
a) Weld metal of passes I or II; b) Parent metal hardened by one pass I or II; c) Parent metal hardened by pass I then by pass II (re-austenitisation); d) Parent metal hardened by pass I and then tempered by pass II; e) Parent metal unaffected.

austenite. So much so that, if during a weld the curve of maximum temperatures reached $\theta_m = f(x)$ with the left hand side of the iron-carbon diagram relative to hypo-eutectoid steels (C < 0.9%), we can interpret as follows (Fig. 6.15) the constitution of a steel with X% of C until maximum temperature is achieved:

a) assuming the parent metal is in the annealed condition, it is made up of ferrite α and pearlite (lamellar aggregate α + Fe_3C at 0.9% carbon) in the proportions indicated by the diagram for the carbon content X%.

b) at temperature A_1, pearlite is transformed directly into austenite of which the carbon content varies by γ_5 (eutectoid carbon content, i.e. 0.9%), from this level, to X%, for temperature A_3 of the steel in question; this mechanism implies the disappearance of the corresponding quantity of ferrite. The result of this is that at a temperature θ between A_1 and A_3, regions of austenite with a carbon content equal to γ_θ (mid-way between γ_s and X) coexist with regions of ferrite which diminish in proportion as θ increases, whilst carbon content γ_θ diminishes, as indicated by line SG of the diagram.

c) when temperature A_3 is reached, the regions of austenite become edge to edge for they have replaced the initial ferrite in attaining the carbon content X%. All traces of the previous structure and constitution are thus eliminated: the welding heat cycle plays the same role as a normalising treatment, in so far as subsequent cooling permits the return to the state of equilibrium.

d) once the austenitic state is achieved from A_3 in the heat affected zone, there is no further solid state evolution, unless of structure, because overheating appears, its intensity depending on the welding procedure.

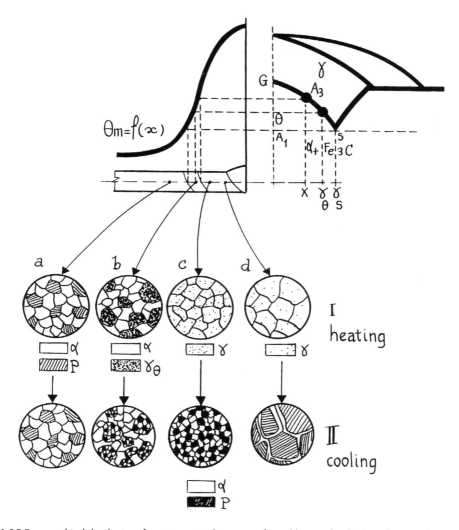

6.15 Topographical distribution of constituents in the vicinity of a weld on steel with X% carbon, in relation to temperatures reached, confronted with equilibrium diagram; resultant structures after cooling involving no hardening: I – where maximum temperature is reached; II – after cooling.

It is from the constitutions and structures b, c and d in Fig. 6.15 that, as soon as cooling commences, the transformations to be examined occur. It should always be noted that these transformations are influenced by the structural state of the austenite on commencement of cooling, this state being influenced by the speed of heating, temperature reached and time at temperature. These factors act on the dissolving of the carbides of added elements and on the homogenisation of the solid austenite solution. This is why in welding we prefer to use specially produced transformation diagrams rather than those which take account of standard heat treatments. We shall come back to this point (in Chapter 7).

Fusion zone

The final stage in the evolution, at increasing temperature, of a metal or alloy is fusion; but whilst this is obtained directly for a pure metal at its particular fusion temperature, for an alloy or simply an impure metal, there is a temperature gap between the beginning and end of fusion, corresponding to what we call the 'solidus'[18] and 'liquidus' of the equilibrium diagram. In this case, from the solidus a liquid phase develops at the grain boundaries existing at high temperature, the proportion of this phase increasing up to complete liquid state. The composition of this phase and its evolution are shown in the corresponding equilibrium diagram (Fig. 6.16). If the heat cycle is rapid, the homogenisation by diffusion required by the equilibrium state is incomplete, which causes the intergranular liquation phenomenon called 'burn', perceptible after cooling, because it gives rise to irreversible heterogeneity; there is at the grain boundaries a phase rich in constituent B, in the case of an initial state made up of a solid solution AB. Intergranular liquation is obviously encouraged by the pre-existence of a fusible phase at grain boundaries, but it is not a necessary condition. In reality, applying this to welding poses the question of knowing why the phenomenon occurs so rarely and so discretely, even for a process as slow as vertical electroslag welding. Here again, the answer, illustrated by Fig. 6.17 lies in the fact that fusion welding involves a moving heat source;[19] at the front of the molten pool, liquation can intervene and a few grains be momentarily surrounded by a liquid phase. The result is decoherence of those grains, which fall and become diluted in the molten pool as that pool progresses. So, after the molten pool has passed, there is no longer any

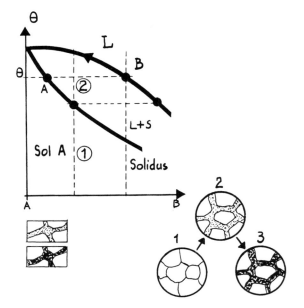

6.16 Intergranular liquation mechanism in a binary alloy: 1 – initial solid solution A; 2 – solid solution A and intergranular liquid B; 3 – condition after cooling: solid solution A and intergranular solid solution B.

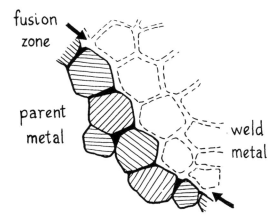

fusion zone

parent metal

weld metal

6.17 Intergranular liquation mechanism at the fusion boundary. The grains torn from the parent metal are represented by the dotted lines.

trace of the grains thus laid bare and we find at the fusion boundary only the grains which, on the molten pool side, have only partially undergone liquation and not been torn away as the molten pool passed by.

The zone concerned is thus very narrow, but it is not without consequence, when the phenomenon occurs, from the point of view of hot cracking or corrosion (stainless or refractory steels) or mechanical properties (oxidised copper). For steels, as we have already mentioned (Chapter 6), this behaviour is associated with decarburisation of the fusion zone.

Chapter 6 **Comments**

1 In effect, this term is also used to described the layout of the atoms which define a crystalline lattice.
2 In reality, the phenomenon starts at a temperature below θ_R , at the crystal scale, by a return to their equilibrium position of the atoms which had been displaced by the slipping and corresponding distortion; we say that there is 'restoration'. This return results in the relief of residual stresses.
3 This difference in behaviour on the basis of the width of the zone affected exists for all the causes of softening which will be mentioned later. This difference will be mentioned again in Chapter 10 with regard to tensile tests on welded test pieces.
4 'Buttering', initially developed by Boutte of the Institut de Soudure for welding self-hardening steels, may be used in difficult cases as a preparatory phase of a welding operation. The filler may be identical to or different from the parent metal or the metal used later for welding. By this surfacing, an intermediate layer is created which causes less stress than direct welding, and which protects the parent metal during subsequent welding. An example has already been given in Chapter 2, with regard to lamellar tearing.
5 This decreases as metal purity increases.
6 Overheated grain size is a function of a parameter associated with the soaking temperature and logarithm of time.

7 Unless, depending on the chemical composition, the transformation $\gamma \rightarrow \delta$ which results in a thin fine grained edging. This constitutes an exceptional case.

8 In particular for welding of chromium ferritic steels, where the purpose of preheating (for example for welding of cast items) is essentially to avoid the risk of brittle fracture due to stresses, by working above the transition temperature.

9 Except if a precipitate plays an embrittling role by 'decorating' the sliding planes.

10 In the case of austenitic stainless steels, we say 'hyperhardening', which distinguishes this case from that of martensitic hardening of steels.

11 Parent metals and filler products must be treated carefully to avoid contamination which would incorporate carbon, via grease for example.

12 Mean composition Al94-95, Cu3.5-4.5, Mn0.5, Fe0.4, Mg0.7, Si0.4-0.5.

13 Except in spot welding, because of the brevity of the heat cycle. The construction of the 'Caravelle' airplane is a classic example.

14 More precisely, ageing results in an increase of transition temperature. Embrittlement may not be apparent at ambient temperature if the transition temperature in the annealed state is low enough.

15 Hence the relevance of the 'accelerated ageing' tests which consist of cold deforming samples (for example at 10%), which are then heated (at 250 °C) and finally subjected to impact tests.

16 In the following text, when we are talking about welding, we say that steel is hardening when it undergoes the hardening phenomenon under the conditions where it is welded.

17 This example has been chosen to facilitate understanding due to the fact that, in the case mentioned, the weld metal is not sensitive to hardening, and because the hardened and tempered zone may be the seat of a particular type of cracking (called heat cracking) which will be commented on in Chapter 9.

18 In a heterogeneous product, it may be a 'false solidus' situated below the true solidus which only exists in the case of complete diffusion.

19 We can however cite a case in spot welding, namely copper enrichment of the periphery of spot welds on Al-Cu alloys which also permit radiographic inspection of the spots (Chapter 10).

7 Solid phase transformations during welding (Cooling)

General points

This chapter, relative to the solid phase transformations which take place during cooling in welding, concerns the heat affected zone of the parent metal as well as the fusion zone during its post-solidification cooling, whether or not that cooling is the first (multipass welding).

Whether they occur in the parent metal or in the fusion zone, these transformations concern mainly the physico-chemical constitution of the material in question, because the micrographic structure (as already defined earlier) which results from heating remains unchanged on cooling or at least this structure determines hereditarily, to a large extent, the structure of the constituents appearing during cooling. So, to study cooling transformations, we need only concern ourselves with the more or less reversible manner in which occur the reactions arising from those reactions which affected constitution during heating, namely for the most part, dissolving into solution without any change of phase and allotropic transformation. During cooling, nothing happens in the zones where heating has not modified the initial constitution.

Two-phase alloys free from allotropic transformation

Return to the state of equilibrium

During our study of heating transformations, we saw that the final stage of heating during welding of a two-phase alloy constituted at ambient temperature and in the state of equilibrium by a phase B heat soluble in a phase A (Fig. 6.9) includes a single solid solution A, which is found in the heat affected zone in the immediate vicinity of the fusion zone. If the cooling which follows is not too quick, the return to the equilibrium state normally takes place, and we see the reappearance of the two-phase state A + B. This situation is most favourable if the parent metal is of the same constitution. Such is the case, for example, for Al Mg alloys, constituted in the equilibrium state, in which they are used, by an aluminium rich solution A, the component Mg_2Al_3 playing the role of constituent B. This constituent dissolves on heating then precipitates out again on cooling, in the heat affected zone and in the weld metal. This is the reason why Al Mg alloys are valued for welding, because, in the micrographic structure (grain size) the heat cycle causes few changes in the affected zone, compared with the parent metal.

It is possible also that achieving the equilibrium state in the HAZ is undesirable, when its appearance replaced another initial state of the parent metal. This case has already been mentioned with regard to duraluminium type alloys,[1] for which the reheated state obtained in the HAZ (following too slow a cooling) corresponds to mechanical characteristics much inferior to those of that alloy in the delivery condition, i.e. hardened and aged.

Two-phase hardening

The term two-phase is used here to distinguish this case of hardening, which concerns the solubility of a phase in another, from that presented by steel, where, as we shall see later, three phases are present (ferrite, austenite and cementite). This two-phase hardening consists of maintaining a solution in phase A, by sufficiently rapid cooling, of a constituent B, which is then in supersaturated solution at ambient temperature. We shall repeat here the examples already given in relation to heating transformations, with some additional information.

In Cr-Ni austenitic steel, the hardened state (we often say 'hyper-hardened') is the normal state for use, either naturally, because of a low carbon content, or by rapid cooling, for a higher carbon content, to avoid chromium carbide precipitation. For all welding processes and carbon contents in current use, the majority of the HAZ of the parent metal and the weld metal remains austenitic during cooling, so that, for single pass welding, the only thing likely to happen on reheating (in a limited temperature field) is precipitation of carbides, which was dealt with in the previous

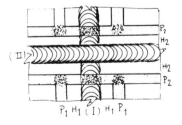

7.1 Zones of hyperhardening on cooling (austenite) and precipitation on heating (austenite and carbide) in the vicinity of crossed weld beads on Cr-Ni austenitic steel: P_1 P_2 zones of precipitation due to the first and second passes respectively; H_1 H_2 hyperhardened zones due to the first and second passes respectively.

chapter. The phenomenon may nevertheless occur in zones hyper-hardened by a first cycle for any welding procedure involving successive cycles, for example in multipass welding, rewelding or back welding, or even crossing of weld beads. So it is (Fig. 7.1) that a corrosion test on crossed beads, carried out on steel susceptible to intergranular corrosion, reveals areas of precipitation in the weld metal of the first bead, and also in the areas of the parent metal hyper-hardened by the heat cycle of that parent metal.

In very high strength alloys, an interesting example is that of 'maraging' steel (mar as in martensite). This complex, nickel based alloy, is characterised by a very low carbon content, such that the martensite obtained[2] by hardening is a soft constituent which, in the virtual absence of carbon can take into solution other elements or constituents which are themselves capable of playing a hardening role if their precipitation is provoked by further treatment after hardening. Welding leads to hardening (soft state) in the HAZ and a subsequent heat treatment (at 480 °C) permits an inverse of hardness. Therefore it is possible either to weld on a hard state (hardened and tempered) and after welding carry out a local rehardening treatment, or to weld on a soft state and then follow up with an overall hardening treatment (Fig. 7.2a and b respectively). In both cases, due to coalescence, there remains outside the affected zone an area permanently softened by the welding heat cycle; we try to keep this area as narrow as possible. It can only disappear by means of a complete re-treatment via re-austenitisation.[3]

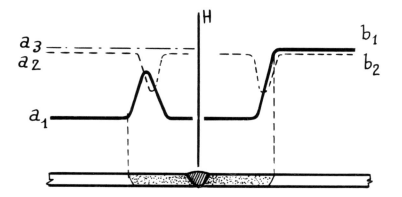

7.2 Hardness relationship in the vicinity of welds on maraging type steel: a_1 initial soft state, as welded; a_2 initial soft state, overall hardening by post-weld treatment; b_1 initial hard state, as welded; b_2 initial hard state, local rehardening by post-weld treatment; a_3 initial soft state, overall treatment of hyper hardening after welding and hardening.

Involvement of an allotropic transformation: steel

General points – study methods

The iron-carbon diagram (already mentioned to give an account of the process of steel transformation on heating – Fig. 6.15), also shows the succession of phenomena during cooling starting from the austenitic state obtained at high temperature, a state which, at equilibrium, is maintained up to the temperature of point A_1: as a result of the progressive ferrite (called pro-eutectoid) separation between A_3 and A_1 which dissolves very little carbon, the austenite becomes rich in this element until it reaches eutectoid content (0.85%) at the temperature of point A_1 at which it gives rise to a ferrite-pearlite assembly constituted of pro-eutectoid ferrite (that which precipitates between A_3 and A_1) and pearlite, lamellar aggregate (α + Fe_3C) at 0.85% C.

But these constituents and their proportions only conform with the indications of the iron-carbon diagram in the equilibrium state, that is for sufficiently slow cooling from a chemically homogeneous and not overheated austenite. Any change in these conditions and in the chemical composition (elements other than carbon) results in divergencies from the information in the iron-carbon diagram; these divergencies can affect the proportion or nature of the constituents as well as their structure.

To take account of the influence of the law of cooling from the austenitic state obtained at a given temperature, continuous cooling transformation (CCT) diagrams (Fig. 7.3)[4] have been plotted in which we find for a given assembly of cooling laws

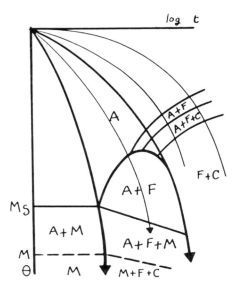

7.3 Diagrammatic example of CCT diagram of a C-Mn steel showing the three conditions for austenite transformation in fields separated by two critical cooling laws (solid line) temperature/time.

from a single austenitisation temperature the commencement, stopping and end of transformation temperatures, together with indications on the nature and proportion of constituents which appear at each moment along the length of each cooling curve.[5]

Such diagrams, which can be used for standard heat treatments, have opened the way to a better understanding of cooling transformations during steel welding. But they have proved quantitatively unusable because they take no account of the influence of two important characteristics which arise in welding, i.e. an austenitisation temperature which is variable but reaching high values (very much higher than in heat treatment) and reached for a short time (very much shorter than in heat treatment). So it was necessary to plot CCT diagrams specially drawn up for welding, directly readable in relation to the welding conditions (energy), themselves often transposed in terms of cooling time corresponding to the various thicknesses and initial temperatures (Fig. 7.4). These diagrams were drawn up either by simulation (heat treatment imitating the welding cycles applied to test pieces) or by in situ analysis, i.e., on actual weld beads or welds, in the affected zone, or in the fused metal on its first cooling. Diagrams of this type are often accompanied by curves indicating the maximum under bead hardness versus cooling times (see Chapter 8).

Whether or not they are drawn up for welding, CCT diagrams show, for each steel, three distinct modes of austenite transformation; for each steel, the appearance of these depends on the law of cooling, namely:

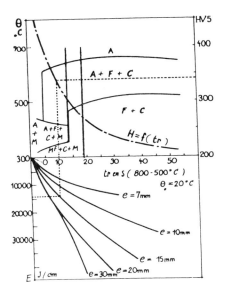

7.4 CCT diagram created for welding, designed in terms of the cooling time corresponding to the welding energy of arc welded beads deposited on metal plates of various thicknesses. This representation is completed by underbead hardness curve H = f(TR) in terms of cooling time. The critical cooling laws are shown here as lines perpendicular to the time axis (critical cooling time).

a) Pearlitic mode (or Ar'), where the constituents which appear are in accordance with the Fe-Fe$_3$C diagram and appear in the order indicated, viz, pro-eutectoid ferrite followed by a ferrite-cementite aggregate resulting from a germination and growth mechanism of the cementite. This mode, which involves diffusion, prevails for slow cooling, i.e. in the right hand section of the CCT diagrams.

b) Bainitic mode (or Ar''), in which the final constitution (ferrite and cementite) is, here also, nearly in accordance with the Fe-Fe$_3$C diagram; but the chronology of transformation is different: the ferrite appears (also by germination and growth) from the austenite, retaining the carbon in supersaturated solution momentarily, then this carbon precipitates in a finely dispersed form. This mode, in which diffusion is also involved, is encouraged by certain alloying elements and occurs for the intermediate cooling laws. It presents several structural variations some of which connect it to pearlitic transformation (upper bainite, acicular ferrite), others to martensitic transformation (lower bainite).

c) Finally, a martensitic mode (or Ar''') results in the direct precipitation of a martensitic constituent, with no diffusion, in which the carbon content is the same as that of the prior austenite, and which appears abruptly, by graduated fractions in relation to the decreasing temperature, from a temperature called M_S (S for starting) down to a lower temperature M_F (F for end (fin)). The martensitic transformation occurs for the fastest cooling laws, that is in the left hand side of the CCT diagrams.

These three transformation modes (for which we shall consider the welding aspects later) occur in fields delineated by 'critical laws'[6] (Fig. 7.3) for the standard diagrams, or 'critical cooling times' (Fig. 7.4) for the welding CCT diagrams. It should however be remembered that a cooling law does not necessarily give rise to a single transformation mode. Depending on the arrangement of the curves, we can have along the length of a single cooling law one, two or even three modes following one another in the scale of decreasing temperatures. Indeed, the pearlitic mode occurs at the highest level (below A_1), the bainitic mode in the middle and the martensitic mode at the lowest level (from M_S). This is the reason why, although the next two paragraphs are devoted to the characteristics of pearlitic and martensitic transformation during welding respectively, we have not attempted to tackle the same subject for bainitic transformation. Indeed, this mode occurs most often, especially in welding, either to complete a transformation started in the pearlitic field, or by preceding a martensitic transformation. In the first case, it is a structural variation of pearlitic transformation (which can be advantageous for the weld metal, as we shall see later). In the second case, coexistence with martensite calls for the same observations and precautions as for martensite.

Characteristics of pearlitic transformation during welding

Single pass welds

When steel is welded under conditions permitting the pearlitic mode to occur during cooling, the appearance of the constituents ferrite and cementite occurs in accordance with a process and leads to a structure which, for a given steel, depends on the initial grain size of the austenite being transformed in addition to the speed of cooling.

If that grain is relatively fine, as is the case around the periphery of the austenitised zone, the diffusion process required by the germination and growth of the cementite can occur unhindered and we see a separation of the pro-eutectoid ferrite, followed by the formation of lamellar pearlite (aggregate Fe α + Fe$_3$C), leading in welds to a grain structure comparable with that of the parent metal (or even finer) if this latter was initially in the annealed state. However, when we look closer we can see that the interlamellar space (distance between the alternating lamellae of Fe α and Fe$_3$C) is smaller in the heat affected zone than in the parent metal which cooled relatively slowly after rolling, for this interlamellar space diminishes when cooling speed increases. The result is moderate increase of hardness in comparison with the parent metal.

If the austenite grain being transformed is coarse, as in the overheated zone, the diffusion process mentioned earlier is hindered and pro-eutectoid ferrite separation tends to occur not only at the boundaries of the austenite grains, but also inside the grains, where it settles along the planes of greatest atomic density, the lamellar aggregate (Fe α + Fe$_3$C) then clogging the area delimited by those planes.[7] Thus a structure is formed (Fig. 7.5b) which is more or less related to the Widmanstaetten

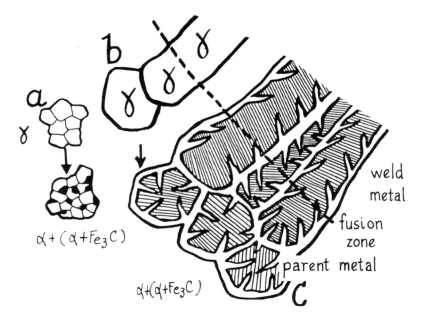

7.5 Conditions of pearlitic transformation in welding: a) Fine austenite grain: granular structure; b) Coarse austenite grain: superheated inter and intragranular ferrite structure. Structures a and b relative to the affected zone of the parent metal correspond respectively to points c and d of Fig. 6.15; c) Austenite grain in the as-solidified state: inter and intragranular precipitation of pro-eutectoid ferrite, orientated from the fusion boundary by epitaxy on the parent metal.

structure,[8] characteristic in the geometric arrangement of the ferrite bands which results from the orientation along the crystallographic planes of the prior austenite. This transformation gives rise to structural heredity, referred to earlier. The limits of the overheated austenite grains which gave rise to this structure are emphasised by the pro-eutectoid ferrite intergranular lattice, whilst the dimension of the intragranular areas is inherited from that of the prior austenite grains.

Transformation in the weld metal never gives rise to a granular structure like that described above even when it occurs in accordance with the pearlitic mode; the orientation of the solidification grains, their size and, possibly, their chemical heterogeneity provide conditions for the appearance of a structure of the type described for the overheated zone, namely the appearance of pro-eutectoid ferrite at the boundaries and interior of the austenite grains. In view of the epitaxy across the fusion zone (already emphasised in connection with solidification), that explains the structure illustrated in Fig. 5.15 and repeated in Fig. 7.5c.[9]

As for mechanical properties, the transformations which occur during the heating then cooling of the heat affected zone cause a variation of mechanical properties compared with those of the parent metal, if the parent metal is in the initial annealed condition; the reduction in the interlamellar space of the pearlitic aggregate is accompanied by a moderate increase in hardness, (and therefore tensile strength and yield strength) and a reduction in notch toughness (or more accurately an increase in transition temperature) in the overheated zone. This relative embrittlement, together with a certain brittleness of the weld zone due to the importance of the pro-eutectoid ferrite bands, becomes more pronounced as overheating is more intense, in connection with higher welding energy like that used in submerged-arc or vertical electroslag welding.

Under these conditions, for application where a notch toughness value is imposed in the overheated or weld zone, we may be constrained to carry out a normalisation heat treatment, i.e. austenitisation at a temperature just above that of point A_3 followed by air cooling. Carried out in this way, this re-austenitisation eliminates the previous structure both in the overheated zone and fusion zone. Another solution making it possible to avoid such heat treatment consists of using the capacity of certain alloying elements to reduce austenite grain size (or the size of the solidification grain in the fusion zone) and/or modify the process of ferrite precipitation during cooling. These two precautions follow the same objective, namely to avoid as far as possible the large-scale precipitation of pro-eutectoid ferrite[10] at the boundaries or interior of austenite grains by replacing it with a finer and more dispersed ferrite, called acicular ferrite, very probably related to the upper bainite resulting from the Ar″ transformation.

Multipass welds

When a fusion weld is made in several passes, the corresponding heat cycle (already studied in Chapter 2) causes re-austenitisation in all or part of the weld metal already deposited and transformed and also in the HAZ, with a consequent new transformation on cooling; this can occur one or more times at a given point, depending on the arrangement and size of the passes, until temperature A_3 is no longer reached. Thus in the case of a three-pass pipe-line type weld (Fig. 7.6) the second pass may totally re-austenitise the weld metal and the zone affected by the first pass,[11] whilst the third pass affects only a part of the weld metal of the second pass. If the steel and welding conditions permit, each re-austenitisation due to a pass gives rise to structural regeneration which results in refinement of the ferrite and pearlite grains resulting from the previous pass. It should also be pointed out that this effect is encouraged in fused metal by its low susceptibility to superheating.

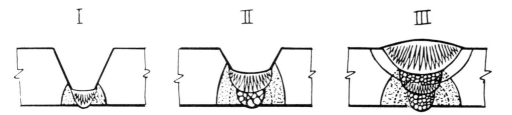

7.6 Structural evolution of the weld metal during a three-pass weld on steel.

The impact of this behaviour on mechanical properties, in particular of the fusion zone, has already been examined with regard to solidification (see Chapter 5); when weld metal is transformed in the pearlitic field, a multipass weld is always tougher than a single pass weld of the same thickness.

Characteristics of martensitic transformation during welding

Transformation chronology and topography

The CCT diagrams show that, for a given steel, martensitic transformation which takes place partially or exclusively from a critical cooling law – i.e. below a critical cooling time in the case of welding, occurs below temperature M_S which is the lowest of all the transformation temperatures able to affect the steel in question. If we return to the time-temperature curves described when we studied the welding heat cycle (Fig. 2.6) we see (Fig. 7.7) that each temperature, in particular M_S, is first achieved on cooling at the periphery of the HAZ in such a way that martensitic transformation progresses from the exterior towards the interior of that zone. On a longitudinal section, the corresponding isotherm, represented by an oblique line, separates the still austenitic metal (in front) from the metal turning martensitic at the rear. This chronology plays an important role with regard to hydrogen embrittlement; indeed, if the hydrogen comes into play from the weld metal, as we stated with regard to development of the corresponding zone (Fig. 4.8) it advances to meet the martensite which is being formed. So, in welding, the problem of hydrogen embrittlement arises in a situation where hydrogen

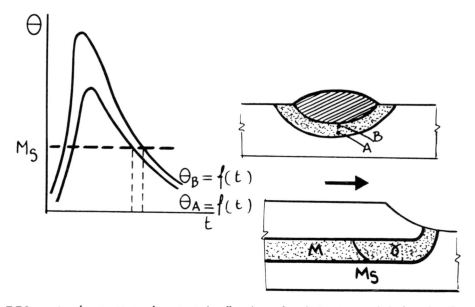

7.7 Progression of martensitic transformation in the affected zone, from the interior towards the fusion boundary (transverse section) and towards the front (longitudinal section).

is already present in the prior austenite at the moment of its transformation into martensite.

Depending on the composition of the weld metal,[12] for a given heat cycle, martensitic transformation may or may not occur. If it does not occur, i.e. if the transformation is pearlitic or bainitic, the isotherm of that transformation is in front of that of the martensitic transformation of the parent metal, which leads to diffusion of the hydrogen across the fusion zone, as illustrated by the diagram in Fig. 4.10; so there is the possibility of embrittlement of the martensite in the HAZ, and therefore of cold cracking. If, on the contrary, the weld metal is more hardening than the parent metal, the corresponding isotherm is in front (situation b of the same figure), the weld metal itself is susceptible to embrittlement and therefore cold cracking. In Chapter 8 we shall return to the problem of cold cracking in steel welding. But for the moment we should note that martensite and hydrogen constitute only two conditions of the phenomenon, a third condition being provided by the presence of stresses.

To complete our examination of the martensitic transformations which can accompany the welding operation, we need to examine the particular case of the zone which, on the outside of the region clearly austenitised by heating above A_3, undergoes a cycle culminating in a temperature between A_1 and A_3. Returning to Fig. 6.15, we can see that for a steel with X% carbon, heating to a temperature θ produces bands of austenite with a γ_θ carbon content between X and γ_S (i.e. eutectoid content) amongst the initial ferrite grains. In the case of sufficiently rapid cooling (low energy welding, spot welding, or even surface hardening), these austenite bands are liable to harden because of their higher carbon content than that of the parent metal. Thus we obtain a mixed ferrite-martensite constitution.

Hardness – brittleness and embrittlement – stresses

Martensite is characterised by a centred quadratic crystalline lattice which is derived from the centred cubic lattice of the ferrite by insertion of carbon atoms (Fig. 7.8) in numbers corresponding to the content of that element which is equal to that of the prior austenite. Consequently, the mesh height, together with hardness and brittleness, increase with carbon content (up to 0.85% C for hardness); also, martensite formation takes place with expansion, hence the appearance of residual stresses which should be taken into account in the balance of all the stresses liable to contribute to cold cracking.

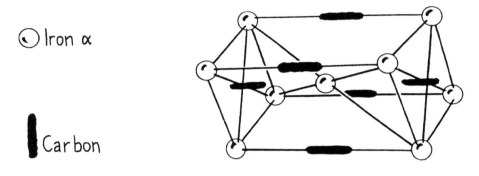

$\bigcirc$ Iron α

$\blacksquare$ Carbon

7.8 Crystalline structure of martensite: insertion of carbon atoms deforming the normally cubic centred iron lattice (quadratic lattice).

With regard to hardness, for a good understanding of hardening in welding and the corresponding precautions, it is important to note that hardening does not only depend on carbon content. In the hardened condition, a non-alloy steel in the martensitic state, given equal carbon content, is as hard as an alloy steel in the same state. On the other hand, the martensitic critical cooling law is influenced by alloying elements so that alloy steel is more hardening. Consequently it will be easier, acting on the cooling law, to avoid hardening by welding for a non-alloy steel than for an alloy steel.[13]

With regard to brittleness, often considered to be systematically associated with the presence of martensite, this is also dependent on carbon content. Low carbon martensites are not brittle, including those resulting from the welding heat cycle, and we even use the term ductile to describe the martensites with less than 0.1% C found, for example, in steels with 0.9% Ni, or to describe the maraging steels mentioned earlier.[14] Such steels easily take martensitic hardening because they are highly alloyed but their martensite is soft because it is low in carbon.

So, in all cases for the carbon contents of steels used in welded construction, the appearance of martensite would cause no serious problems if it were not for the fact that that constituent is made brittle by hydrogen, in particular when that gas is already present in the austenite lattice which produced the martensite as is the case in welding. When hydrogen is included in its lattice, martensite which is not very brittle in itself becomes embrittled and may thereby contribute to the cold cracking mechanism. Of course, the susceptibility of martensite to hydrogen embrittlement depends on carbon content; when carbon content increases, hydrogen tolerance decreases, so that welding processes characterised by a low hydrogen content in the weld metal (for example MIG with uncovered wire or even TIG) pose problems in welding with high strength and high carbon steels, whilst those processes are considered safe in this respect for lower strength constructional steels.

Tempering during post-weld treatment

As we have seen with regard to heating transformations, martensite, an unbalanced constituent, tends to break down into ferrite and cementite under the effect of reheating to a temperature lower than that of A_1. This is the tempering phenomenon. Beyond A_1, re-austenitisation occurs, and all traces of previous hardening disappear. Having already pointed out the effects of this phenomenon during multipass welding (Chapter 5), it remains for us to mention it in relation to the post-weld heat treatment which we call stress relieving. In effect, taking account of the temperature range currently practised, always lower than A_1, there is no stress relieving without tempering. Thus, any zone of an assembly which is still martensitic (or tempered at low temperature during a multipass weld) undergoes tempering during a stress relieving treatment. The resulting effect is in principle beneficial if tempering alone is involved. But other phenomena may intervene in the sense of deterioration, whether in the form of embrittling precipitations or the type of cracking called reheat cracking. Both aspects will be included in the chapter dealing with heat treatments (Chapter 9), together with the hardening treatment followed by tempering, practised on welded elements or assemblies.

Chapter 7 **Comments**

1 See notes 12 and 13 of Chapter 6.
2 For this type of alloy, it is the martensite which plays the role of solid solution A mentioned earlier with regard to the hardening constituents which play the role of B.
3 As for the weld metal, the choice of a suitable grade makes it possible to compensate for the loss of titanium which occurs during welding.
4 In French TRC (Transformation en Refroidissement Continu.)
5 The CCT curves, which effectively permit such forecasts, should not be confused with the TTT curves (time, temperature, transformation) which do not permit them, because they concern austenite transformations in the isotherm condition.
6 The transformation temperatures in each range and the position of the laws or critical times are influenced by carbon contents and addition element contents.
7 The lamellar aggregate is no longer eutectoid but hypo–eutectoid. We sometimes talk of 'pseudo–pearlite'.
8 This general term designates the structure observed when a phase separates inside the

phase grains which produced it and orientates itself on the crystallographic planes of that prior phase.

9 It should be remembered that the austenite grain of the weld metal does not necessarily coincide with the grain produced by solidification if that solidification took place in the delta field. But the epitaxy across the fusion boundary and the elongated form are nevertheless observed for austenite.

10 The gross character of the ferrite accentuates the susceptibility of the ferrite to by cleavage, itself encouraged by its centred cubic crystalline lattice.

11 Because of this effect this second pass is sometimes called the 'hot pass'. Its influence is the same regardless of the transformation (pearlitic or martensitic) resulting from the first pass.

12 It has been established that, with identical chemical composition, the weld metal is less hardening during its first cooling than the parent metal; this is explained by its structure and especially by the role of inclusions.

13 We shall return to these respective roles of carbon and alloying elements in connection with the 'carbon equivalent' notion (Chapter 8).

14 For these latter steels, it is precisely the rarity of carbon atoms in the lattice which permits a structural hardening mechanism by means of other elements.

8 Hardening and cold cracking in steel welding

General points

Cold cracking in steel welding, associated with the three characteristic features martensite, hydrogen and stresses, has already been mentioned in previous chapters where these factors were studied separately.[1] Given the importance of the phenomenon, which dominates the problem of steel weldability and for that reason has been very much involved in the evolution of steels for welded construction, it is important that we should return to it in a summary chapter devoted to hardening during welding and its main risk, cold cracking. In particular this chapter includes an analysis of the criteria used and a metallurgical justification of the recommended welding procedures.

The term cold cracking is used here in its most general sense and, as we shall see later, it covers several types and morphologies.[2] We talk about *cold* cracking because it results from embrittlement of the martensite which itself appears at low temperature in the scale of transformations.[3] We understand that this embrittlement is caused by hydrogen, at least in French terminology; English literature specifies the role of this gas by using the expression 'hydrogen induced cracking'.

Hardening factors during welding

Nature of the steel - notion of carbon equivalent

As we have already said, the tendency of a steel to hardening is reflected by the position of its critical martensitic cooling law in the time scale. In practice, with welding, hardenability in the heat affected zone can be evaluated by the value of the critical cooling time, itself dependent for a given thickness on initial temperature and energy used. We have already emphasised the respective roles of carbon content (martensite hardness) and addition elements (hardenability).

Although these respective roles of carbon and addition elements are differentiated, it has for a long time been the practice to group these elements together to calculate, on the basis of the chemical composition of a steel, a number likely to take account of that steel's performance from the point of view of welding induced hardness. This number, called the *carbon equivalent*, is obtained by adding to the carbon content the alloying element contents, each being assigned a coefficient taking account of the role played by the element in question with regard to the criterion chosen, for example hardness increase in the heat affected zone.[4] This is how we often find the following so called IIW formula in literature:

$$C_{eq} = C + \frac{Mn}{6} + \frac{Cr + Mo + V}{5} + \frac{Cu + Ni}{15}$$

In practice when comparing steels, given the importance of manganese, we often merely calculate the value of $C + Mn/6$. This is what we have done below to explain how, despite its apparently rule-of-thumb character, this formula is likely to give the user better information than the simple enumeration of chemical composition. Indeed, the specification of the maximum carbon equivalent, thus defined, complements that of the carbon maximum giving a better location of the curve defining maximum underbead hardness (the use of this will be explained later) on the basis of cooling time;[5] if we consider two steels A and B one of which has maximum carbon content (therefore minimum manganese content) and the other having minimum carbon content (therefore maximum manganese content), their hardness/cooling time curves cross (Fig. 8.1). This is because steel A has a higher intrinsic hardness than steel B, but a lower hardening capacity. If therefore we require a welding condition – and a cooling time – ensuring a given maximum hardness H_{max}, we shall find that time in the area framing the two curves A and B. Thus, the carbon

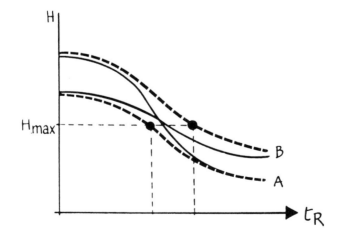

8.1 Interpretation of underbead hardness criterion in connection with carbon equivalent: A cooling time/hardness curve of a steel rich in carbon and poor in manganese; B cooling time/hardness curve of a steel of the same carbon equivalent, poor in carbon, therefore rich in manganese. The broken lines indicate the limits defined by the two extreme examples.

equivalent specification completes that of the maximum carbon, permitting a valid correlation between the underbead hardness and cooling time. We need to remember that this correlation can only be safely used, from the point of view of weldability, for the purposes of comparison between steels of the same type, i.e. containing the same alloying elements in different proportions. For example, for C-Mn steels, a maximum value of 0.40 (for C + Mn/6) can give the user useful information on the necessity of particular precautions if that value is exceeded.

In this way, the notion of carbon equivalent has played an important guiding role in the perfecting of welding steel procedures and also the development of steels for welded constructions.

Welding heat cycle

The possibility of the welding induced hardening phenomenon depends directly on the welding heat cycle, via the factors already mentioned, namely: component mass, energy and initial temperature (preheating). But the existence of steels which, by their chemical composition, systematically become hardened when welded, has led to the search not only for welding procedures which avoid hardening, when that is possible, but also those which, whilst not avoiding hardening do reduce the risks, in particular from the point of view of cold cracking. This is why post-heating is also included in the present discussion, even if it is practised at a temperature which does not modify the transformation on cooling.

Component mass

As regards the effect of component mass, we can usefully refer to Chapter 2 concerning the thermal study of welding. We need only add two comments, one thermomechanical, the other metallurgical.

From the thermomechanical point of view, we must emphasise that, even if the notion of limit thickness is involved in fixing, at a constant value from a certain thickness, the cooling time associated with a given energy, the same is not true for the stress condition generated by the welding operation, which in all cases of multipass welding becomes more intense as thickness increases. Consequently, martensitic transformation is accompanied by the risk of cold cracking, this risk becoming greater when thickness increases.

From the metallurgical point of view, we must remember that the product standards generally specify higher carbon and/or alloying element contents for thick products than for thin products. Thick products are, therefore, more hardening because of their chemical composition (at equal strength) than thin products, hence an additional risk of cold cracking with thick products.

Welding energy

For steels not systematically hardening during welding, the hardening incidence is determined by the welding energy, in association with initial temperature, where the welding energy results in a cooling time longer or shorter than the critical martensitic time. In this respect, the use of CCT diagrams is valuable, or failing that, it is possible to use hardness criteria (to which we shall return later). In practice, it should be remembered that low welding energies give rapid cooling and the greatest care should therefore be taken where such energy is involved, for example:

– the first passes of multipass welds which are the most severe in terms of the heat cycle;

– tack welds which should be avoided or carried out with the same precautions (preheating) as the actual welds;

– erection welds, the effects of which are often neglected, especially as they are destroyed after use, without being sure of the disappearance of a possible underlying crack;

– welds, even unstressed, securing various items (stops, projections);

– arc initiations outside actual welds; these cause intensely hard craters which may include cracks and detract from mechanical performance (fatigue), even if there is no fused metal deposit. Such practice is moreover often prohibited.

Initial temperature: preheating

Preheating, practised under the conditions and with the precautions indicated in Chapter 2, serves as increased energy and contributes to increased cooling time and thus the avoidance or reduction of martensitic transformation by exceeding the corresponding critical martensitic cooling time. But the role of preheating goes further for it can be effective against cold cracking, even if the martensitic transformation takes place. So we must now consider all the effects of preheating[6] in steel welding not only with regard to martensitic transformation, but also from the point of view of stress and hydrogen evolution (Fig. 8.2) along heat cycles without preheating (solid line) and with preheating (dotted line).

With regard to martensitic transformation, this can be avoided by means of increased cooling time if the steel lends itself to this in its CCT diagram but, even if martensitic transformation takes place, it occurs between M_S and M_F in a time which is increased by preheating (shaded zone on the curves). The transformation is thus less abrupt.

As for the welding stresses which, as we have seen, are set up during and up to the end of cooling, their accession to the maximum value is delayed by preheating through increased cooling time (measured here by the cooling time between maximum temperature and 100 °C).

As for the hydrogen present, which scarcely had time to diffuse during cooling without preheating (hence the final H_0 content) this has the chance to do so at least partially, (hence the final H_p content). The result is that, even if martensitic transformation does take place, cold cracking can be avoided if, at the moment when stress σ_p is produced, the hydrogen content in the weld has reached a sufficiently low H_p value. In the absence of preheating, stress σ_0 is produced earlier, whilst hydrogen content H_0 has not developed much, hence a possibility of cracking.

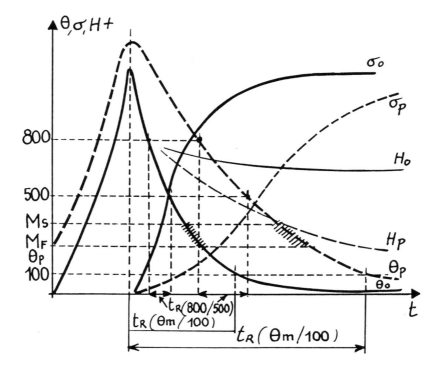

8.2 Comparative evolution of temperatures, stresses and hydrogen during welding without preheating (solid line, index 0) and with preheating (broken line, index p). The symbols t_R 800/500 and t_R θ_m/100 designate the cooling time between 800 and 500 °C and maximum temperature and 100 °C respectively.

Thus, depending on the nature of the steel, the role of preheating holds good either vis-a-vis the appearance of martensite, evolution of hydrogen and stresses, or solely on these last two factors.[7] In the first case, there is no risk of cold cracking; in the second, the risk is reduced.

Postheating

In the text which follows, the term postheating is exclusively reserved for the operation already described which consists of interrupting cooling for a given period at a given temperature before allowing cooling to continue. This practice should not be confused with heat treatment after welding (or post-welding) which consists of reheating the welded joint or part after complete cooling. The most usual practice is to fix a postheating temperature equal to that of the preheating, but it is also possible to operate at a higher or lower temperature. When postheating is performed at a temperature lower than M_F, which is most frequently the case, in addition to making the temperature uniform in the weld zone, it also gives rise to evolution of hydrogen and stresses which complete and extend the influence of the preheating, with no metallurgical modification since all the martensite which could form has already appeared (Fig. 8.3).

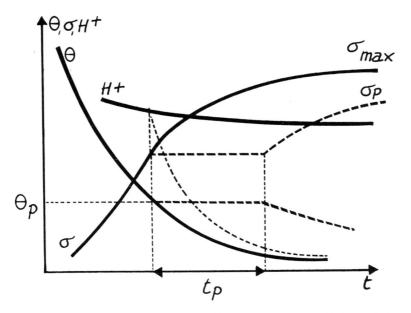

8.3 Comparative evolution, from maximum temperature, of temperature, stresses and hydrogen during welding with preheat alone (solid line) and with preheating complemented by post-heating (broken line).

With regard to hydrogen, postheating provides this gas with time for diffusion on the basis of temperature θ_p at which it takes place and time t_p during which it lasts. As for stresses, whose value depends on temperature, they cease to grow during time t_p, and only begin to grow again when postheating ceases. So their maximum value (on complete recooling) is delayed until hydrogen diffusion has reduced the content of this gas to a harmless value; cold cracking is thus avoided, provided that temperature and postheating time have been well chosen. Raising the temperature reduces the time necessary so that the use of postheating makes it possible to reduce the preheat temperature,[8] for a given situation in matters of hydrogen content, stress condition and welding energy. The representation in Fig. 8.4 shows an example of how time t_p necessary for non-cracking varies in relation to temperature θ_P. Thus designed and practised, postheating has made it possible to extend considerably the possibilities of using high tensile steels and also to improve operator comfort, because of the reduction in preheat temperature.[9]

Consequences of hardening during welding

Hardness increase – underbead hardness criterion

As we have seen earlier, martensite hardness depends on its carbon content; also, in all cases for steels liable to escape hardening (i.e. self-hardening), the occurrence of martensite in the heat affected zone and in the weld

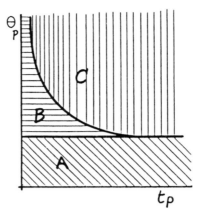

8.4 Fields and circumstances of cracking according to the associated values of temperatures θ and postheating times t_p (all other parameters identical): A cracking during postheating: insufficient temperature; B cracking after postheating: insufficient time; C no cracking: correct association of temperature and time.

metal zone depends on the cooling conditions which characterise the welding procedure used. This is why welding induced hardening, even if it has no quantitative consequences on mechanical performance,[10] but being relatively easily measured, is often chosen as a criterion of approval or testing of the welding procedure, under the name of underbead hardness.

Underbead hardness, in particular that measured in the heat affected zone surrounding a weld bead deposited in a single pass on the surface of a product of a given thickness is the maximum value – or rather, depending on the relevant standard*, the mean of the two maximum values – found in this zone by applying the procedure described in that standard. This criterion permits a simple determination of the welding conditions providing a maximum hardness value below the required value. Thus (Fig. 8.5) for a given steel and thickness, the line of underbead hardness/energy at various initial temperatures makes it possible to choose the most favourable energy/initial temperature combination ensuring the required maximum underbead hardness: for example, if θ_0 equals ambient temperature (absence of preheating), it will be necessary to provide energy at least equal to E_0, whilst E_1 will be sufficient if preheat to temperature θ_1 is used. One or other of these two combinations may be selected depending on the welding conditions: site or workshop, welding position, operator's comfort, etc.

With regard to the specified maximum hardness value, we cannot ignore the value $H_{max} = 350$ Vickers, often cited, even if it calls for some comment; at the time, already long past, where the 350 Vickers value has been put forward in a rule of thumb manner (in connection with the notion of equivalent carbon, as we have said

---

* NF A 81–460: Steel products. Methods of determining underbead hardness.

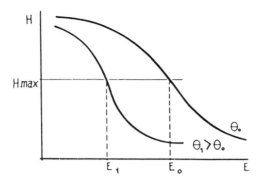

8.5 Influence of initial temperature on the position of hardness/welding energy curves.

earlier), the role of martensite in the cold cracking process appeared much more important than that of hydrogen, which was ignored for a long time. Also, for steels used in metal construction, given their carbon content, a hardness of 350 HV seemed to be the limit value revealing the presence of a proportion of martensite giving rise to a risk of cold cracking. Consequently, limiting maximum hardness to this value boiled down to forbidding martensite, and thus those welding procedures which led to its formation. With increasing knowledge on the role of hydrogen and processes and filler products permitting lower contents of this gas in weld metal, there was a gradual rise in the value of acceptable maximum hardness, which meant that a certain proportion of martensite was permissable. So values between 350 and 400 were specified, and even relationships between acceptable maximum hardness and carbon content were proposed.[11] This evolution was more justified with the perfecting of welding procedures and the generalised use of low alloy steels, for which the traditional idea of maximum underbead hardness had lost its significance.

Brittleness – hydrogen embrittlement

As we have already said, martensite brittleness depends on carbon content, so that, for steels habitually used for welding, brittleness is not dangerously involved, especially as the tempering produced by the superimposed heat cycles of multipass welding reduces the effects of hardening.

As for hydrogen embrittlement which is the basis of cold cracking, in addition to what we have said previously we should add that it is not a case of acquired embrittlement; indeed, it only appears when hydrogen is present and ceases as soon as that gas disappears by diffusion. Hence the interest of the precautions just described, such as postheating and, to a lesser extent, preheating. In the same way, as embrittlement becomes more significant with increasing carbon content, so the time during which it may be apparent after welding is longer (delayed cracking).

Fatigue performance

Although martensite, through its susceptibility to notch sensitivity (itself dependent on carbon content) may be intrinsically subject to fatigue cracking, for example in the case of a uniform component, its occurrence during welding most often results from a mechanical rather than metallurgical influence. In effect, its presence in a welded joint (weld metal or HAZ) creates a 'metallurgical notch'* effect (by the localised hardness increase it produces) which exists within the confines of the hardened zone, by reason of the discontinuity of the corresponding yield strength compared with that of the parent metal. It is at this spot that fatigue fractures are most likely to be initiated in components subjected to cyclical stresses. It is the frequency of incidents caused by this mechanism, in mechanical components, which justifies the care (already emphasised) which needs to be given to designing these components and to carrying out the welding, by seeking to avoid low energy welds, in particular because of their small dimensions, even when these welds do not transmit the stresses themselves.

Of course, in addition to this metallurgical notch sensitivity effect there is that of residual stresses in so far as these have not already contributed to a process of cold cracking. It goes without saying that a pre-existing crack situated in a zone already prone to fatigue can only be conducive to fracture initiation. So may appear the necessity for heat treatment, either stress relieving (which thus procures a tempering effect) or even complete hardening and tempering.

Risk associated with hardening: cold cracking

Description of the phenomenon

Cold cracking occurs when the three conditions for its appearance, hardening, hydrogen and stresses are combined. This is the reason why it occurs at the end of cooling, sometimes after an appreciable time following complete cooling (delayed cracking).[12] The more or less tardy nature of cold cracking results from that fact that a certain delay occurs before the necessary quantity of hydrogen collects, by diffusion under the influence of stresses, at a given point. Hence the occurrence of privileged sites, such as the toe or root of weld beads, due to the concentration of stresses created by the notch they constitute.

From the morphological point of view, cold cracks (Fig. 8.6) can be classified on the basis of their position (heat affected zone, fusion boundary, weld metal zone) and their orientation (longitudinal or transverse). Below are a few comments on the formation conditions of these various types of cracks and the names given to some of them.

---

* H de LEIRIS: Fatigue and welding – Soudage et Techniques Connexes, 1963, no 7/8, 253–266

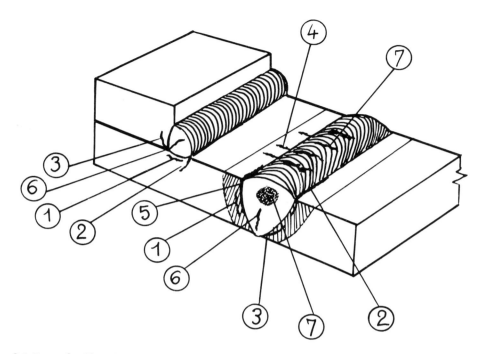

8.6 Types of cold cracking affecting fusion welds on steel, classified on the basis of their position and orientation (single pass butt weld and fillet weld).

HAZ cracks

Cracks in the HAZ are most often longitudinal. It is in the HAZ that we find 'underbead' cracks[1] which, in the main, appear parallel to the fusion boundary and have the peculiarity of not emerging on the surface. They appear with a heavy degree of hardening and high hydrogen content, which makes cracking possible, even with low stress, as is the case for a single bead deposited on the surface of a metal sheet or plate (hence their name).

Two other types of longitudinal cracks, 'toe cracks'[2] and 'root cracks'[3] occur with a relatively low hydrogen content because they arise from a notch sensitivity effect at which stresses are concentrated and gather the hydrogen necessary for cracking, which can thereby be deferred, whilst underbead cracking is produced as soon as cooling is complete.

Transverse cracking[4] occurs more rarely in the heat affected zone, in the presence of relatively high carbon martensite, when the assembly configuration lends itself to a predominance of longitudinal stresses, for example, in an external fillet weld on thick plate.[13]

Fusion boundary cracks

In the fusion boundary, cold cracks[5] , longitudinal by definition, may occur in heterogeneous hardening steel welds with austenitic filler metal, because of a heterogeneous dilution (not taken into account in Schaeffler's representation – see Chapter 4) which results in the formation of a martensitic strip in the weld metal, along the length of the fusion boundary and very susceptible to hydrogen embrittlement.

Cracks in weld metal

Cold cracking occurs in weld metal when it is as hardening or, as already shown (Chapter 7) more hardening than the parent metal. Cold cracking is either longitudinal[6] due to the notch sensitivity effect at the root, or transverse[7] , emerging or not at the bead surface. This latter type of cracking is due to a mechanism related to that of fish eyes (see Chapter 4). In effect, cracks initiate from defects (inclusions or porosities) which gather embrittling hydrogen, with the help of longitudinal stresses.

Multipass welds

In multipass welds, the conditions governing cold cracking do not inevitably occur with each pass. A multiple pass weld can be completed without renewed cracking if the phenomenon is produced, for example, between the first and second passes; but the renewed fusion caused by the second pass may partially modify the appearance of earlier cracks without removing them (Fig. 8.7) whilst the corresponding heat cycle transforms the surrounding structure. We thus find cracks situated in annealed zones whilst they result from previous hardening. In another context, we can also see cold cracks appearing at a weld pass (Fig. 8.8) because of stresses due to shrinkage of other passes, for example on the reverse side of a butt weld.

Cold crack detection

The harmful role which cracks can play with regard to service performance, and more particularly cold cracks because of their fineness, implies particular vigilance during non-destructive testing of welds and finished welded assemblies.

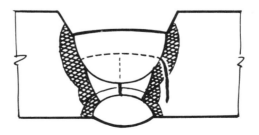

8.7 Modification by a subsequent pass of the appearance of cold cracks caused by a previous pass.

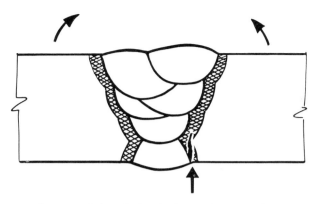

8.8 Example of a toe crack generated along the length of a pass by stresses due to subsequent passes.

With regard to the time when inspection should be carried out, two recommendations can be made, namely:
– since cracking may be delayed, inspection should not be carried out too soon after completion of welding, especially if stresses are applied soon after welding (erection/fitting, tests,[14] etc)
– if the tested assembly is to be heat treated, for example stress relieved, testing should be carried out *after* treatment, because the cracks are made less fine and therefore more easily detectable, and also to take account of their evolution during the treatment, for example by reheat cracking.[15]

Given the variety of types of cold cracks, it is clear that the various test procedures are not all equally satisfactory depending on the position and orientation of cracks whose presence may be suspected by means of crack tests carried out when establishing the welding programme. If we look again at the diagram in Fig. 8.6, we see for example that magnetic particle inspection is suitable only for cracks which emerge at or close to the surface (type ②, ⑥ or ⑦ under certain conditions[16]). Radiographic inspection detects only the cracks into which the radiation penetrates ②, ⑤, ⑥, provided they are not too fine. Finally, because these are flat defects, only ultrasonic inspection can detect all types of cold cracks, in so far as a good understanding of their possible position and orientation permit the operator to take the necessary steps for that detection.

As for metallographic examination, its destructive nature obviously excludes its systematic use – but in exceptional cases, it may be extremely useful in identifying certain cracks and avoiding confusion (for example ② and ⑤ which have the same external appearance). So this examination is carried out on samples guided by the results of non-destructive testing (see Chapter 10).

Cold cracking tests

The numerous cold cracking tests suggested by the litera-ture differ in the manner in which the restraint necessary for cracking is carried out, or the stress resulting from that restraint is obtained. As examples of restrained weld tests, we will mention here two internationally-used tests, the British CTS test (controlled thermal severity) and the TEKKEN test developed in Japan. These two tests call for self-restrained welds, i.e. where the restraint is obtained by the arrangement of the components concerned.

The CTS test (Fig. 8.9) is performed on two plates of the steel to be tested, first bolted tightly against each other, then joined by two fixing welds F_1 and F_2. Test fillet welds C_1 and C_2 include a more or less severe heat regime, for a given welding energy, depending on associated thicknesses and joint geometry, which may be 'bi-thermal' or 'tri-thermal' (see Chapter 2). The cracks, sought on transverse bead sec-tions, may be located in the toe or root of the weld metal.

The TEKKEN test (Fig. 8.10) includes the depositing of a test bead on a butt weld of two chamfered pieces of the steel for testing, partially assembled by welding at each end. The form of the chamfer may be modified to vary the conditions at the toe and root, the most severe form being the Y with a slight gap. Cracks are observed mainly in the root, either in the affected zone of the parent metal, or in the weld metal zone.

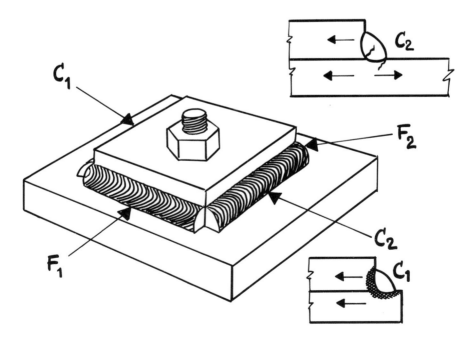

8.9 Controlled thermal severity test (CTS): F_1 F_2 fixing beads; C_1 C_2 test welds (C_1 bithermal – C_2 trithermal).

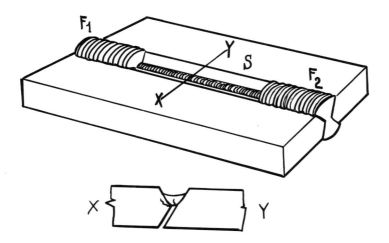

8.10 TEKKEN test principle: F_1 F_2 fixing beads; C_1 C_2 test welds, examined on XY section.

The results of the CTS test may be made relatively quantitative by the notion of thermal severity it proposes. As for the TEKKEN test which, as carried out, appears rule-of-thumb, it has in reality been the subject of much work establishing a correlation of its results with measurements of critical crack stress on a restrained weld made by means of an instrumented set-up (RRC test) derived from that illustrated in Fig. 2.25 relative to the definition of restraint intensity (Chapter 2).

It was the wish to make a quantitative evaluation of cold cracking susceptibility which led the Institut de Soudure to develop the implant test. The fundamental principle of an implant, already mentioned during our study of the welding heat cycle and used for the in situ determination of CCT diagrams, was used to make a crack test testpiece from the implant itself. To that end, it is notched[17] and a specially adapted mechanical rig used to apply a constant tensile load, after deposit of the experimental weld bead, which may cause cold cracking at the notch after a given time (Fig. 8.11a). By means of a series of experiments involving various cooling times, it is possible to plot a crack stress curve[18] on which we can read the cooling times necessary to guarantee no cracking for a given stress, for example the yield strength or any other lower value (Fig. 8.11b). To make this choice possible, the Institut de Soudure, at the same time as compiling CCT and implant cold crack diagrams* drew up a series of crack stress values to be used for different types of assembly, in relation to their restraint intensity.

---

* IS/CETIM: Practical documentation on weldability of steels – Institut de Soudure – Cetim – 1976.

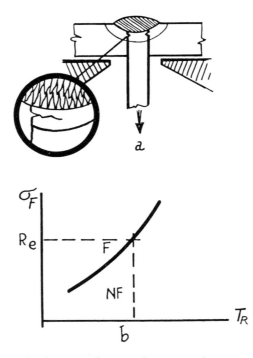

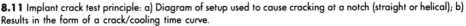

8.11 Implant crack test principle: a) Diagram of setup used to cause cracking at a notch (straight or helical); b) Results in the form of a crack/cooling time curve.

The implant crack test is covered by two French standards[**] and one recommendation from the International Welding Institute[***]. It has inspired numerous studies on the phenomenon itself and is widely used for the determination or verification of welding procedures. The same test technique makes it possible to evaluate the cold cracking tendency of the metal deposited by a filler product (covered electrode, wire-flux couple or cored filler wire).

Practical conclusions - miscellaneous precautions

Most of the practical indications given by the specifications or the recommended practices for welding of steels take account essentially of the risk of cold cracking. For metal construction steels or boilers and pressure vessels, of the C-Mn-Si type, with or without dispersoid elements, these recommendations do not specify particular tests, unless it be an underbead hardness testing. Based on adequate knowledge of the performance of those steels, itself the result of systematic

[**] NF A89–100: Implant cold cracking test methods
NF A03–185 Conventional methods of characterising cold cracking weldability of weldable steels by the crack test on helical notched implants
[***] IIS/IIW-447-73: Recommendations for the use of the implant test as an additional information test on cold cracking susceptibility when welding steels - Welding in the World, 1/2, 1974, pp 9 - 16.

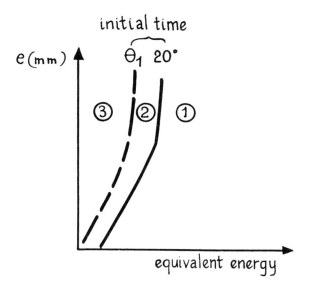

8.12 Diagrammatic example of 'weldability range' as proposed in French standard A36-000: 1 Welding without preheating (20 °C); 2 Welding with preheating (θ_1); 3 Welding with post-heating.

tests, these recommendations suggest 'weldability fields' (Fig. 8.12), as does the section in French Standard A36-000* for example; these recommendations indicate the 'equivalent energy'[19] to be used, in relation to thicknesses to be welded, in order to obtain a fixed percentage (85% as indicated above) of the maximum hardness read on the 'hardness-criterion' curve for the type of steel used.

In alloyed steels, for which (as already stated) underbead hardness cannot be taken as a criterion of weldability, welding procedures must be established on the basis of cold crack test results taking account of the degree of restraint. The collection already mentioned meets this objective.

However the problem of cold cracking is approached, it is necessary to quantify restraint or, more generally the stress exerted on welded joints during welding and, if possible, to limit it, at the risk of exaggerating the precautions to be taken, in terms of energy and preheat temperature. These can be harmful not only from the metallurgical point of view (needlessly coarse and therefore less tough structure) but also from the practical point of view (welding position incompatible with high energy, operator discomfort). In this respect, although we now have data on the restraint corresponding to various assembly geometries, we should not lose sight of the causes, of stresses other than restraint stresses to which a welded joint may be subject during welding, such as:

---

* NF A36-000: Recommendations on the suitability for welding of metal construction steels and for boilers and pressure vessels

- stresses caused by making a neighbouring weld or one on the reverse side of the welded product,
- stresses caused by the welded part's own weight (e.g. welding on the tension flange of a girder),
- stresses caused by the reaction of a part deformed elastically to permit welding (ferruling, attachment of stiffeners, etc).

In such cases, parts should be secured so as to prevent these stresses having an influence before the hydrogen is dispersed by diffusion.

Evolution of steels for welded construction in relation to weldability requirements

The problem

From the beginning of welded construction, we have seen in the technical literature the affirmation that 'carbon is the welder's enemy'. Indeed, whether it is a matter of embrittlement due to overheating or that which, aggravated by hydrogen, is the cause of cold cracking, weldability problems become greater when steel carbon content increases, i.e. when the yield and tensile strength of the product in question rise. In effect, this correlation between strength/carbon content/weldability is inescapable if we hold to what is indicated in the iron-carbon diagram, since, for a given state (most usually the annealed state), an increase in strength is obtained at the price of a rise in carbon content and therefore a worsening of weldability problems. As welded construction developed and specifications became more and more demanding with regard to strength and ductility, it was necessary for steel producers to find solutions which were compatible both with those specifications and with acceptable weldability characteristics. Hence the evolution concerning chemical composition and product delivery condition, the solutions proposed being inspired by one of these approaches, or both.

Evolution of chemical composition[20]

Historically, the search for products which, at the same strength, had a lower carbon content than products in which only carbon was involved, resulted in the appearance of C-Mn-Si steels. For such steels, delivered in the normalised condition (i.e. air cooled after rolling), the influence of the elements Mn and Si is exerted as both the austenite transformation on cooling, and by the strength of the constituents so that the addition of these elements in moderate quantities makes it possible, at equal strength, to lower the carbon content. But the influence of manganese on the hardenability of these steels is revealed also when they are welded, which was the reason for the factor assigned to that element in the formulation of equivalent carbon (for example $C + Mn/6$), so that the maximum manganese content remained fixed at around 1.2%.

To bypass the obstacle now encountered, it was necessary to involve a mechanism for reinforcing the steel which partially replaced that permitted by the Fe-C diagram, whilst not having any significant effect on hardenability. A solution was found in the form of 'dispersoid' or 'micro-alloy' steels which contain very small proportions of addition elements (Al, Nb, V, Ti) of which niobium is the most used, at contents of a few hundredths per cent. The action mechanism of these elements is explained by their affinity for carbon or nitrogen, with which they form components of the carbide, nitride or carbonitride type, components which are insoluble in austenite below 900 °C, and also in ferrite. This variation in solubility makes possible a ferrite hardening mechanism similar to those which were described earlier (see Chapter 6), hence a gain in yield strength, whilst the quantity of pearlite is reduced compared with that found in ordinary steel, for which all the carbon participates in the austenite transformation on cooling. At the same time, this mechanism contributes to a refining of the grain, especially if the product has been obtained by controlled rolling (see later) which results in improved ductility. So, at equal yield strength, the presence of dispersoid elements permits a lower carbon content and therefore better weldability, illustrated by the hardness/cooling time curves or even, at equal carbon content, a higher yield strength, and without maximum hardness being affected; we see only a slight increase in hardening capacity (Fig. 8.13).

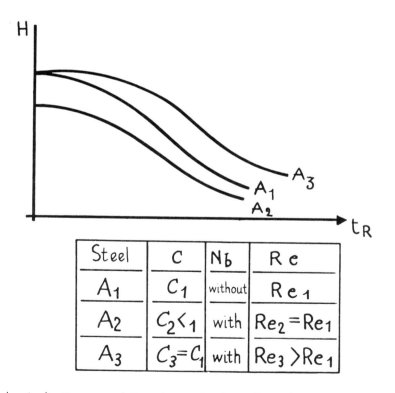

Steel	C	Nb	Re
A_1	C_1	without	Re_1
A_2	$C_2 < _1$	with	$Re_2 = Re_1$
A_3	$C_3 = C_1$	with	$Re_3 > Re_1$

8.13 Hardness/cooling time curves relative to three steels with or without niobium and comparative levels of yield strength.

Product delivery condition

The correlation between the carbon content of a steel and its tensile characteristics is valid only for the constitution indicated by the Fe-C diagram, i.e. for the state of equilibrium, as obtained by annealing or normalising, or at least by hot rolling. If, therefore, we wish to increase tensile characteristics without increasing still further the carbon content and without further additions, another possibility consists of modifying the condition in which the products will be used so that their characteristics can be guaranteed.

Controlled rolling

A first alternative, which modifies only the structure of the products and not the constitution, concerns the method of rolling; it has been named controlled rolling. This technique consists of doing all or part of the hot rolling, not at high temperature, but at a temperature as close as possible to the temperature at which austenite transformation begins on cooling or even in the range of temperature where that transformation occurs.[21] Under these conditions, a much finer grain than by standard high temperature rolling is obtained because the recrystallisation grain undergoes a very much reduced growth. This treatment also encourages hardening by precipitation of dispersoid components, which are finer and better dispersed. Finally, this treatment results in a significant improvement in tensile and especially ductile characteristics, without any deterioration in weldability.[22]

Hardening and tempering

Another solution consists of deliberately deviating from the structural condition (lamellar pearlite and ferrite) implied by the correlation between tensile characteristics and carbon content, by subjecting the products to a hardening treatment following by tempering. In effect, this treatment results in ductile and tensile characteristics which, at equal carbon, permit not only much better performances than in the annealed condition, but also make it possible to reduce carbon contents for equal performance. We can see immediately the advantages offered by this possibility, from the point of view of weldability, with regard to both cold cracking and ductility, even if welding of hardened and tempered steels requires certain precautions – already mentioned in Chapter 6, in particular to limit the extent of the over-temper, by using minimum welding energy and initial temperature.

For relatively thin thicknesses (up to about 22 mm) the preparation of hardened and tempered products poses no difficulty from the point of view of hardening penetration. Hardening of the full thickness, after rolling or integrated in the rolling equipment[23] can be achieved without special additions. Products thus treated, characterised by a low carbon content[24] vis-a-vis their tensile characteristics, pose no problems of weldability other than that of over-tempering by welding, which in this case is easy to resolve and, moreover, the hardening and tempering treatment makes it possible to obtain excellent ductility characteristics. A current spectacular application of this is welding of high tensile steel oil or gas pipelines working at low temperature. On the other hand, for thicker products it is not possible to obtain full thickness hardening without the addition of elements such as Ni, Cr or Mo to modify the critical hardening law, but then that produces a steel which, despite its low carbon content, hardens when welded; this must be taken into account when establishing welding procedures. This can be delicate because the desire to avoid cold cracking leads operators to increase initial temperature and welding energy, whilst the pre-occupation concerning over-tempered softening calls for reductions in these values.

Finally, whatever the means employed to obtain the hardened and tempered state, all post-welding heat treatment at a temperature higher than the temperature of the initial temper undergone by the product must be prohibited. Indeed, such a treatment would produce an over-tempered induced softening of the assembly which could be unacceptable.

Dual-phase or reduced pearlite steels

Finally, we can mention variants which are even more sophisticated, combining several of the solutions already mentioned and achieving very much improved weldability because of a very low carbon content (C < 0.10%) associated with excellent ductility. They are in the main 'dual-phase' steels hardened between A_1 and A_3, or 'reduced pearlite' where the bands of lamellar pearlite are replaced by bainitic bands. In both cases, the effect of a trace element (Nb) is reinforced by that of a hardening element (Mo for example).

All of these solutions are summarised by the representation in Fig. 8.14, in the form of the association they make it possible to obtain between yield strength and ductility*.

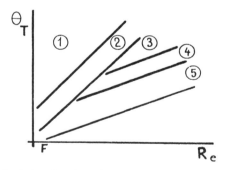

8.14 Influence of state of rolled steel on the yield strength/ductility association (according to Irvine): R_e yield strength; θ_T impact strength transition temperature
1 fine grain pearlite and ferrite
2 pearlite and ferrite – controlled rolling
3 bainite
4 bainite – controlled rolling
5 martensitic hardening and tempering

---

* K J IRVINE: ISI Publications, no. 104, 1967

Influence on filler products

The evolution of steels for welded construction which has just been described has led to a distinct improvement in weldability by reducing the risk of cold cracking in the HAZ whilst at the same time improving ductile and tensile characteristics.

As for filler products, a parallel evolution has occurred, in every case for hydrogen content of the deposited metal, concerning the manufacture of those products (in particular covered electrodes), and by reason of wider use of procedures depositing metal with lower hydrogen content, such as MIG or MAG used with bare wire. But obtaining high tensile deposit metal in the as-deposited state in one or more passes necessarily implies the use, in addition to carbon, of alloying elements making it possible to obtain in that state mechanical characteristics equivalent to that of modern parent metals. Because of this, the problem of cold cracking now concerns weld metal as much, if not more, than parent metal. Fortunately, the precautions we need to take are the same and everything we have just said remains valid.

Chapter 8 **Comments**

1 Whilst cold cracking only occurs if these three factors intervene by their combined effects.

2 We often talk about 'underbead crack'; in fact, this term concerns only one of these types.

3 The cracking which occurs when ammonia is present also takes place cold. But it has a progressive character and, for that reason, is related to stress corrosion. Moreover, the cracks are rarified, which is never the case for cold cracks.

4 Such seems initially to have been the case. Subsequently, numerous formulae were proposed, some of which also take account of hydrogen content together with degree of restraint in evaluating the risk of cracking. The following comments rest solely on the criterion of increase of hardness.

5 In France, such a curve has been named 'hardness/criterion' curve.

6 Apart from the simple effect of drying of welded parts, which constitutes an additional advantage of the preheating.

7 When it does not avoid the martensitic transformation, preheating in no way modifies hardness in the heat affected zone. For steels which act in this way, the hardness test is not therefore a verification of the efficiency of preheating and the maximum underbead hardness criterion loses its significance.

8 We are talking here of components of average importance because, as we have already observed, the overall preheating of a heavy component is equivalent to prolonged postheating, because return to ambient temperature takes so long.

9 More rarely practised, postheating between M_F and M_S or above M_S permits respectively an interruption of the martensitic transformation or a non-martensitic isothermal transformation more or less completed depending on duration.

10 For example vis-a-vis fatigue strength (see later), or even during machining.

11 For example $H_{max} = 240 + 790\ C$ (French standard A36-000 to which we shall return later), which equals at least 80% martensite and corresponds to 382 Vickers for a steel with 0.18% carbon.

12 This delay may be used to advantage when welding in multiple passes to establish a procedure permitting avoidance of the phenomenon (see later).

13 We recall here the comment already made concerning metallographic examination; only a longitudinal section can reveal transverse cracks with certainty. At most, a transverse section may reveal a part of them (for example crack ⑦, but gives a false appearance).

14 We can cite in this respect an accident during the course of the premature hydraulic test; the stresses due to the test caused delayed cracking which was itself the cause of an abrupt fracture.

15 Weld repair of cracked zones may pose a problem with regard to the post-repair heat treatment or even the new cracking which may result.

16 A crack of type ⑤ is, despite appearances, likely to cause confusion when it appears at the fusion boundary of a heterogeneous weld because it may be confused with a toe crack ②.

17 The notch, which may be circular or spiral, permits hydrogen to collect tending to imitate the situation which leads to cracking at the toe or root. In the absence of a notch, we obtain higher cracking stress, corresponding to the underbead cracking, with greater dispersion.

18 We can also plot a fracture stress curve, recommended as a criterion by some authors, by using small diameter implants.

19 The equivalent energy is equal to the nominal energy affected by a factor characterising the geometry of the joint in question (see Chapter 2).

20 With regard to chemical composition, some authors have commented that the search for a very low purity rate leads to rarification of sulphurous inclusions; lacking a sufficient number of inclusions to 'trap' the hydrogen and render it harmless, a steel which is too pure would be more susceptible to cold cracking than an impure steel. Given the advantages conferred by purity from the point of view of mechanical performance, this has not led to practical applications.

21 If we examine the TTT curves, we see that pearlitic transformation is slow, which allows enough time for the operation.

22 The disadvantages lie particularly in production, because of the waiting time necessary for cooling to rolling temperature and also because of the necessary power of the rolling mills since, at operating temperature the steel is much less ductile than at high temperature. So the maximum product thickness benefiting from this technique is limited (of the order of 20 to 30 mm).

23 'Roll quench' arrangement.

24 As far as it is possible to combine the effect of this treatment with that of dispersoid addition elements.

9 Heat treatments for steel welds

Purpose and use of heat treatments

Classification according to purpose

We have seen in preceding chapters that welding results in local metallurgical modifications of the parent metal and, depending on processes and welding procedures, in the creation of a fusion zone whose structure and properties are more or less different from those of the parent metal. Also, welding sets up a state of residual stresses whose distribution and scale are also dependent on the processes and procedures used.

Whether they are metallurgical or thermomechanical, these modifications are often tolerable with regard to service performance and welded assemblies are used in the as-welded state. In certain cases, however, it is deemed preferable or necessary, or even obligatory, to use a postweld heat treatment of which the principal aim may be to improve or obtain those properties considered important for service conditions, or to eliminate (or reduce) residual stresses. In reality, given the multiple roles played by every heating and cooling cycle, the first type of treatment, designed to play a metallurgical role, acts also on the stress state. The second, which is mechanical, nevertheless has metallurgical consequences. So it is by bearing in mind this interaction that we shall study one after the other the following types of treatments, for steels, without going into the nature of the energy sources used:
– stress relieving and tempering treatments;
– normalising treatments;
– hardening and tempering treatments.

Methods of use

We can classify and describe as follows (Fig. 9.1) the way in which heat treatments concerning welds or welded assemblies are practised.

A) The treatment may be total, i.e. affect the constituent weld or welds of the assembly at the same time as the whole of the parent metal. Such an action may be simultaneous (A_1), i.e. the entire welded assembly is subject to the same treatment cycle at the same time, or the treatment may take place step by step (A_2), but for each position it occupies, the heat source acts simultaneously and in accordance with the same cycle on the weld and the unaffected parent metal. As an example of simultaneous global treatment (A_1), Fig. 9.1 illustrates the case of annealing of liquid

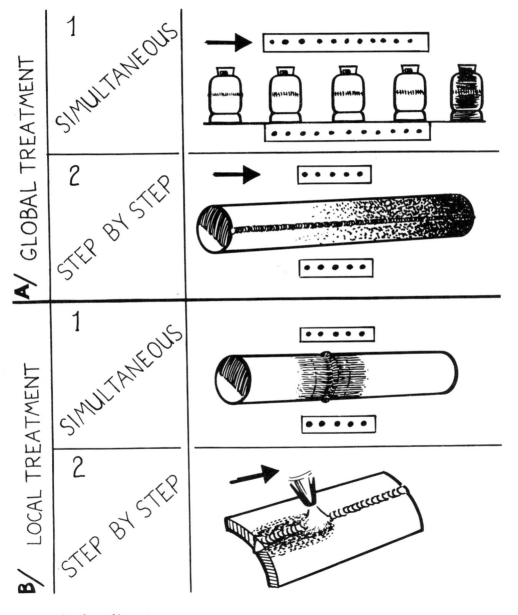

9.1 Modes of use of heat treatments.

gas bottles which, after welding, pass through a tunnel furnace.[1] The step by step global treatment (A_2) is illustrated by a diagram taken from the manufacture of resistance or induction welded tubes; the unmoving heat source treats an entire section of the moving tube, both the parent metal and the newly performed weld.

In practice, simultaneous global treatment can be applied to all the above-mentioned treatment objectives and all shapes and sizes of welded assemblies compatible with the capacities of the heat sources used. The step by step global treatment, which requires a constant product profile, is in the main practised for tube manufacture, for normalising or hyperhardening, depending on whether it is 'black' or stainless steel.

B) Simultaneous local treatment (B_1) consists of operating at the same time on all of a weld and its surroundings and thus on only a part of the parent metal around the affected zone. The treatment is distinguished from local step by step treatment (B_2) in the sense that the latter is practised by means of a heat source[2] moving in relation to the weld or vice-versa (one stationary, the other moving), to provide the step by step effect sought in the weld metal and the part of the parent metal which includes the affected zone.

Simultaneous local treatment[3] is in the main designed for circular welds for which it provides a metallurgical effect (normalisation or tempering), together with stress relief, by reason of the symmetry it ensures. Such is not the case with step by step local treatment which, if it provides a metallurgical action (essentially normalisation) is not recommended to obtain stress relief since it only replaces one state of weld-induced stress by another, induced by the treatment itself.

With regard to local heat treatments, whether simultaneous or step by step, these can only be practised without disadvantages under two conditions, both relative to the parent metal, to which we shall return later.

a) During normalisation treatments, cooling speed and hence moving speed of the heat source, must be regulated so that the treated zone, which is momentarily austenitised, does not harden.

b) Any local treatment creates its own heat affected zone which replaces the HAZ of the weld being treated. Due to the inevitable temperature gradient existing there, this zone includes on its edges a band where the treatment applied has not been complete but where it has created a modification of the parent metal compared with its initial condition, in particular softening if the parent metal itself is in the initial hardened and tempered condition.

Stress relief and/or tempering

Purpose

Stress relieving is used to remove – or diminish – residual stresses present in a welded assembly when it is complete, or during construction if it appears necessary to intervene with regard to stress condition before continuing welding operations in total safety. Given the temperature range in which stress relief treatment is carried out (all cases lower than the temperature of point A_1), it also gives rise to a tempering effect when the welds contain hardening constituents, both in the weld zone and the affected zone. This effect is sometimes the main purpose of the treatment which is still carried out in accordance with the same techniques. Unfortunately, this additional metallurgical role is not always beneficial because unfavourable effects can be produced, affecting mechanical properties or causing reheat cracking .

Stress relieving mechanism

In our thermal and thermomechanical study of welding (Chapter 2) we indicated that thermal stress relieving is obtained because of the reduction of yield strength caused by the rise in temperature; the result of this reduction is that every area of a component thus reheated which contains residual stress at ambient temperature (stress which is obviously elastic) becomes plastic as heating progresses thus providing the elongation necessary to obtain a relief rate which increases with treatment temperature.

A simplified representation of this mechanism can be obtained by comparing on a single graph (Fig. 9.2) the treatment heat cycle $\theta = f(t)$, variation $R_e = f(\theta)$ of yield

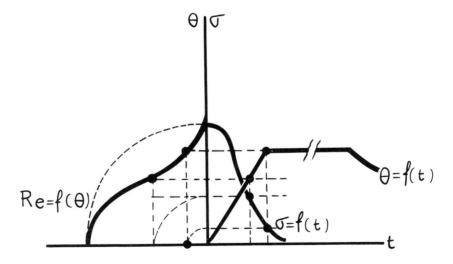

9.2 Interpretation of the stress relief effect (curve $\sigma = f(t)$) obtained during heat treatment (curve $\theta = f(t)$) due to the variation of yield strength (curve $R_e = f(\theta)$).

strength in relation to temperature and finally curve σ = f(t) giving the variation of residual stress during the treatment cycle.

According to the documentation in standard A36-200*, the rate of relief, on first consideration, a function of treatment temperature alone, is very low up to 200 °C, reaches 50 to 60% at 500 °C, 70 to 85% at 575 °C and 90 to 95% for a temperature of 625 °C, the result being achieved as soon as maximum temperature is reached. But it is still true that any heat treatment must be specified not only in terms of temperature, but also in terms of duration, to take account (in relation to energy available to provide the treatment) of product mass treated and the necessity of homogenising temperature to avoid the creation of new cooling stresses. Also, holding the temperature specified for a certain time ensures an additional percentage of relief in the regions where stress peaks exist initially.

Finally, to define a heat treatment in terms of temperature and time, and to evaluate its effects from the point of view of stresses and especially of mechanical properties, the Hollomon parameter is used.

$$H = T(20 + \log t) \, 10^3$$

where T is the treatment temperature (in degrees K) and t is treatment duration (in hours).

This parameter is used in the above-mentioned Standard to take account of the effect of the treatment on mechanical properties, summarised in the next paragraph. To this effect the standard states the H values (Fig. 9.3) corresponding to the various temperature/time combinations.

Metallurgical effects

The heat cycle of a stress relieving treatment being comparable with that of a tempering treatment, there is not on the whole any effect on the parent metal when the parent metal itself is used in the normalised and tempered or hardened and tempered state, except if the treatment takes place at a temperature higher than that of the initial tempering; to that end we generally keep to a lower temperature of around 30 °C less. It should however be pointed out that some alloyed steels (Mn-Mo, Mn-Ni or even Cr-Mo) may be susceptible to temper brittleness which occurs during a treatment at 450 – 550 °C or during cooling in the course of too slow a passage through this temperature gap. This phenomenon, which seems to be associated with the effect of impurities, is less sensitive when vanadium is present.

On the other hand, for steels used in the normalised condition, stress relieving heat treatment acts both on tensile characteristics (at ambient temperature and when

---

* French standard A36-200: Influence of post-weld heat treatments on the mechanical properties of steels for boilers and pressure vessels

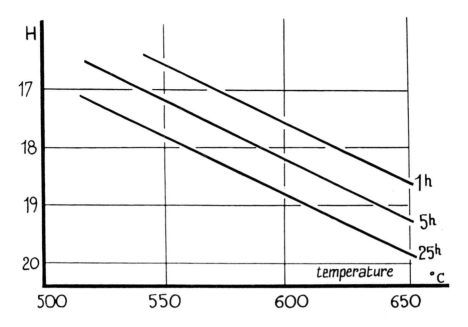

9.3 Chart for the determination of the Hollomon parameter (H) for various temperature and stress thermal relieving treatment time combinations (according to standard A36-200).

hot) and on impact toughness transition temperature; this action depends on parameter H of the treatment, defined as indicated above. This behaviour occurs for C, C-Mn, C-Mn + Ni steels and for dispersoid high yield strength steels. For example, for C-Mn steel, standard A 36-200 (Fig. 9.4) gives the order of magnitude of reduction in yield strength and increase in transition temperature which can be expected in the parent metal, for the various H parameter values.

With regard to the effect of stress relief treatment on the mechanical properties of the weld metal, it is beneficial as regards recovery annealing from the condition aged by the tenso-thermal effect which can be encountered in multipass welds on thick products (see Chapter 6). On the other hand, it can have an embrittling effect for certain compositions susceptible to temper brittleness, mainly during cooling which is too slow; the maximum temperature reached and soaking time are not involved in this phenomenon.

In conclusion, it appears that in any case, we must when making the calculations for a construction to be subjected to stress relief, take account of the actual mechanical properties as they result from the metallurgical effect of that treatment. To do this, value H of the parameter of that treatment must be set at the minimum compatible with the technological conditions mentioned above, and with the rate of relief required or imposed, whilst the cooling speed must be controlled at the maximum possible, whilst maintaining thermal uniformity of the components treated.

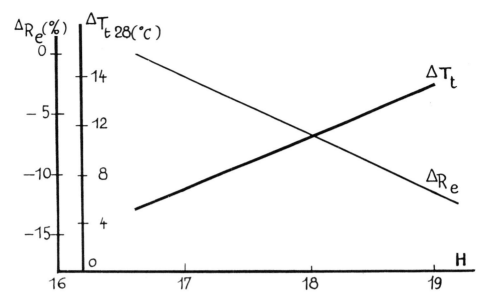

9.4 C-Mn steel: variation of yield strength (ΔR_e) and impact toughness transition temperature (ΔT_t for 28 kJ) according to standard A 36 200, by the effect of thermal stress relieving treatment.

Reheat cracking

The simplified explanation of the stress relief mechanism given above presupposes that the stress and structural states are homogeneous. In reality, a welded assembly, in the as-welded state, is not homogeneous from any of these points of view; the welded joint and its surroundings do not have the same structure as the parent metal and therefore react differently vis-a-vis the reheating necessary for relieving stresses and the stress state is not uniform because of that dissimilarity of structure and also of the geometrical characteristics of the construction. For these two reasons, the welds may not stand up well to the plastic elongation necessary for stress relief during the reheat cycle because of the initial structure there and the modifications that this cycle generates there. This risk is increased in the zones where stress peaks appear. If such is the case, the plastic elongation imposed may be obtained only at the risk of reheat cracking.

Reheat cracking may affect not only welded assemblies, but also surfaced components, in particular interior walls of pressure vessels lined with a layer of austenitic steel deposited by wide pass submerged-arc welding especially with sheet filler metal. In this case, cracking, which is transverse, occurs in the narrow tempered zone of the parent metal between two neighbouring passes, already mentioned with regard to solid phase transformation during heating (see Fig. 6.14). This case poses an additional problem associated with the difference in expansion factor between the austenitic deposit and the parent metal, which occurs during both welding and subsequent heat treatment.

This phenomenon, which does not arise for most steels, in particular C-Mn or dispersoid high yield strength steel, is characteristic of some alloy steels containing alloying elements with an affinity for carbon such as, for example, Cr-Mo-V steels. Metallographic examination of cracks shows that they initiate readily in the affected zone, at the toe of weld beads, where the notch effect locally accentuates initial residual stress and spreads along the prior austenite grain boundaries, which during post-weld cooling had given rise to martensite or bainite (Fig. 9.5). This crack run, which occurs at increasing temperature, is explained by the precipitation of carbides in the prior austenite grains, hence a hardening which leads to localisation of deformation at the boundaries of those grains, weakened by the segregation of the impurities produced there. The risk of reheat cracking is thus mainly associated with the chemical composition of the steel (alloying elements and impurities) and, on the other hand, with the conditions under which the treatment is carried out.

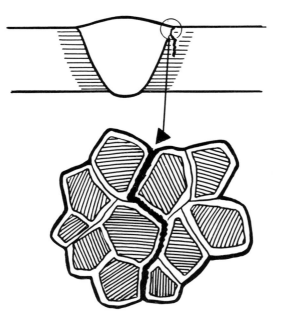

9.5 Macro and micrographic appearance of reheat cracking, affecting the grain joints of the prior austenite of the heat affected zone.

Treatment before or after cooling

Normally, there is no disadvantage in carrying out stress relieving treatment on a welded assembly straight after completion of the final weld, before complete cooling, even if the welding programme specifies postheating.

In effect, the stress relief treatment is carried out at a temperature higher than the normal postheating temperatures and it can only complete the effects. However, if there are several identical components for welding and treating, it may be desirable to store them after cooling and subject them later and simultaneously to the specified heat treatment. But to do this in complete safety, it is necessary to ensure that the postheating time and temperature permit full cooling without risk of cold cracking.

Likewise, when a lengthy welding operation has to be interrupted, for example, for testing purposes, the preheat temperature must be maintained during a sufficient time to ensure cooling without risk, or intermediate stress relief treatment should be carried out. Another solution consists of carrying out non-destructive testing when hot, but this implies some not insignificant technical difficulties.

Treatment including austenitisation

Annealing - normalisation

With steel, the purpose of both annealing[4] and normalising is to obtain, by means of heating above point A_3 and causing complete austenisation of the zone or component treated, followed by relatively slow cooling, a refined structure constituted of ferrite and pearlite which replaces the constitution and structure in the as-welded state. Normalisation, which includes faster cooling than annealing, produces pearlite with smaller interlamellar spacing and therefore harder than annealing. For some low alloy steels, normalisation can produce bainitic type hardening and thereby necessitate subsequent tempering treatment.

From the point of view of mode of use, the simultaneous global treatment is the most complete because it affects the whole of the treated component with no temperature gradient. It therefore uniformly produces the desired metallurgical effect both in the heat affected fusion zone and in the parent metal, which benefits from the same heat regeneration whether it has been deformed cold or at high temperature (hot or cold stamping). If it is a large and heavy assembly, the global treatment may pose difficult practical problems of deformation or bulk. It may be necessary to support components to prevent them being deformed and, for components too big for normalising in one operation (such as long ferrules), it is possible to treat half at a time, putting one half into the furnace whilst the other half protrudes. It is also possible to normalise separately each ferrule element, in which the longitudinal weld, placed vertically in the furnace benefits from the treatment without risk of deformation; then the ferrules are assembled and we finish up with an overall stress relief treatment, without normalisation of circular welds.

The step by step treatment (type A_1 of Fig. 9.1), when it is possible, calls for no reservations except that it is not applicable to a steel which would harden under the effect of the relatively rapid cooling which is characteristic of that steel.

Simultaneous local treatment (B$_1$) does not pose this problem because cooling can be regulated to compensate for the effect of the temperature gradient. On the other hand, on either side of the treated region this gradient inevitably results in the creation of a partial austenitisation zone between A$_1$ and A$_3$ where softening occurs, more or less pronounced depending on the initial state of the parent metal.[5]

The step by step local treatment can only be used for steel welds not liable to harden during subsequent cooling. Also, it is accompanied by residual stresses which means that its use is not advised for applications where there is a risk of stress corrosion.

Spot welding

We have already mentioned during our study of the welding heat cycle (Chapter 2) the possibility of subjecting a resistance spot weld to heat treatment produced by the joule effect of further current on the machine itself, immediately after the spot weld has been made. From the metallurgical point of view there are two comments to be made about this treatment concerning the dead time (tm on Fig. 2.16) separating the welding itself from the subsequent heating, and the cooling law which follows the passage of the welding reheating and current.

With regard to dead time tm, this must be such that the austenite transformation during the cooling which follows the spot welding has enough time to take place, i.e. is longer than the end of transformation time t_F (Fig. 9.6). Otherwise, that is to say if tm is less than beginning of transformation time t_D, the passage of current and the resultant joule effect will only reheat the austenite as it is, without any regenerative effect (cycle II of the same figure).

Because of the high speed of cooling which follows a spot weld, austenite transformation can be deferred to a relatively low temperature, which increases the necessary tm time and constitutes a timing problem in using high strength steels in mass production, in the car industry for example. Without going as far as plotting transformation curves in continuous cooling (difficult), it is recommended that the minimum value necessary for dead time tm be established by tests (for example tension tests – see Chapter 10).

As for the cooling which follows the passage of the reheat current, it is usually sufficiently rapid for there to be no pearlitic transformation, even for non-alloy mild steel so that we cannot really talk of normalising treatment. The effectiveness of the treatment lies essentially in a structural effect due to the passage through the austenitic state, which removes the previous structure of the spot weld and makes it possible to obtain a fine structure on cooling, regardless of the constituents resulting from the cooling law which follows.

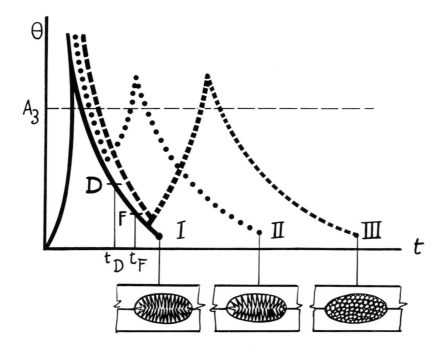

9.6 Heat treatment of resistance spot welds: influence of the dead time between the welding cycle and heat cycle: I welding with no treatment; II premature treatment, occurring before transformation commencement time t_D; III efficient treatment, occurring after transformation completion time t_F.

Intercritical treatment

The term intercritical is used to describe a treatment given at a temperature located between points A_1 and A_3 of the steel in question. According to the iron-carbon diagram, steel in this temperature range is made up of a portion of the initial ferrite, i.e. that which existed prior to the treatment, and austenite resulting from transformation of the initial lamellar pearlitic aggregate.[6] If we start from an out of balance state, the intercritical treatment will also cause the steel to evolve towards this constitution. At the same time, the carbides or carbonitrides present (dispersoid steels) undergo a partial dissolution. On cooling (necessarily slow because we are still dealing with a global treatment) the austenite bands transform into ferrite and pearlite, new and finer, compared with the initial constituents they are replacing and the ferrite bands not dissolved during heating remain intact. It is the same for the behaviour of the carbides and carbonitrides. This mechanism, briefly described, is accompanied by an improvement in fracture toughness (lowering of transition temperature) which, without being as important as that which achieves normalisation is nevertheless significant enough to be interesting, particularly as at the temperature practised, the risk of collapse of components treated is very much reduced.

With regard to welded assemblies, intercritical treatment can be used either in addition to normalising treatment, the effects of which it completes, or as sole treatment of as-welded products, mainly to improve fracture toughness in the weld zone. The further removed from the equilibrium state the initial structure is, the more significant this improvement will be, taking into account chemical composition.

Finally, we must point out that, despite the advantages it brings, the application of intercritical treatment is limited to welded assemblies made up of parent steel used in the normalised state. Indeed, in view of the temperature at which it has to be carried out, necessarily higher than the temperature of point A_1, this treatment cannot be applied to assemblies comprising hardened and tempered steel which cannot, at the risk of unacceptable softening, be raised to such a temperature.

Hardening and tempering of welded assemblies

Although of relatively infrequent application, the hardening and tempering treatment of welded components, which occurs especially in aeronautical or mechanical construction, is worth mentioning here because of the characteristics it presents. The aim of the treatment is to remove the effects caused by welding and obtain, by hardening and tempering, optimal qualities in the welded components, both in the parent metal (which can be welded in the initial annealed condition or already hardened and tempered) and in the weld metal.

In a steel destined for this type of treatment, its composition (carbon and alloying elements) is necessarily that of a hardening or even self-hardening steel, which creates an initial difficulty, during the actual welding, from the point of view of brittleness generated by the hardening due to the heat cycle and also from the point of view of the risk of cold cracking, both these risks being accentuated by a high carbon content, necessary to obtain high characteristics in the treated condition. In particular, cold cracking occurs even with a very low hydrogen content, as found in the weld metal of TIG or MIG bare wire welding.[7]

Also, in so far as these difficulties can be resolved by the precautions indicated with regard to hardening and cold cracking (see Chapter 8), the success of the treatment requires a 'response' to this treatment in the weld metal identical or comparable with that of the parent metal, so that there is no discontinuity at the weld or welds. It is clear that this anxiety leads us to raise carbon and alloy element content in the weld metal to a level comparable with those of the parent metal, but this solution increases the risks of embrittlement and cold cracking already mentioned.

So we have a difficult problem which for each case necessitates prior experiment, including direct tests on welds, or even comparative tests of hardenability of parent metal and weld metal, for example by means of an adaptation of the Jominy test, as described in the Air Standard no. 9117. The Jominy test consists of using a jet of water to harden one end of an austenitised round bar and, after complete cooling, plotting a hardness/distance from hardened end curve. The variant[8] described by the standard consists of prior building up, with the filler product to be tested, of a machined section of the bar. After welding, austenitising and end hardening, the hardness/distance from parent metal and weld metal curves are plotted and respective hardnesses can then be compared (Fig. 9.7).

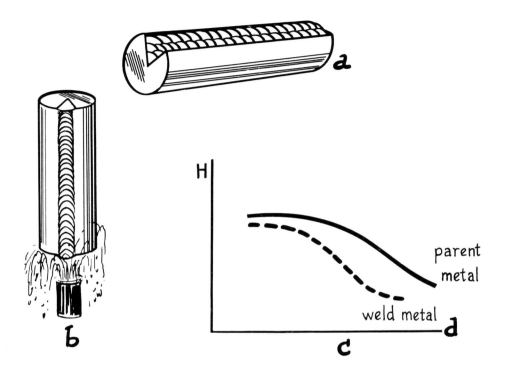

9.7 Utilisation of the Jominy test for comparison of hardenability of weld metal and parent metal: a) Building up on the bar; b) End hardening of the surfaced and austenitic bar; c) Comparison of hardness/distance curves.

Chapter 9 **Comments**

1 This example concerns a case of series production. It is a large assembly, the treatment taking place in a furnace where the entire component, immobile, is raised to the required temperature (or first one half, then the other, after turning).

2 When the heat source is a flame, this operation is sometimes called flight annealing (in French recuit à la volée).

3 In resistance welds (butt or spot) a simultaneous local treatment is carried out, for metallurgical reasons, most often using a reheat current on the machine itself.

4 In the terms of French Standard A02-010 (Heat treatment terminology), annealing, from the austenitic condition, comprises furnace and therefore slow cooling, whilst normalisation comprises air and therefore relatively more rapid cooling.

5 This softened zone is larger than that which normally borders the heat affected zone, because the temperature gradient of the treated zone is much lower.

6 Just like in the zone between A_1 and A_3 heated by the steel welding heat cycle (see Chapter 6, Fig. 6.15).
7 It was the total of the problems thus posed which led the space industry to replace hardening steel of the 40 CDV 20 type (0.40% C, 5% Cr with Mo and V) by maraging steel, mentioned in Chapter 8.
8 This variant was initially proposed by T Noren (Sweden) (reference unknown).

10 Metallurgical aspects of destructive and non-destructive weld tests

General points

The object of this chapter is not to supply information on techniques to be used for weld tests. It is proposed only, as indicated in the title, to comment on the characteristics of these techniques which are dictated by the metallurgical consequences of the welding operation and draw attention to what happens from the point of view of interpreting results.

With regard to destructive tests (limited here to mechanical tests on test specimens) which are intended either to contribute to evaluation of quality, or to supply quantitative data concerning mechanical performance of assemblies tested; the test techniques used are derived from the traditional techniques described in the various standards and regulations. But these techniques must be adapted and the results interpreted because, created initially for application to homogeneous products (or in all cases to products in which heterogeneity constitutes an anomaly), they are used in the case of welding on test specimens which are congenitally heterogeneous. Indeed, independent of any discontinuity of shape or form, a welded test specimen contains weld or heat affected zones in which local properties are different from those of the parent metal, which is also present. For each test category, it is therefore necessary to examine the consequences of this heterogeneity, especially as there may also be accidental and transient heterogeneity caused by possible defects.

For non-destructive tests, the role of structural or constitutional heterogeneity, although less systematic, must also be examined, with regard to the possibilities of identification of defects. But, from the historic point of view, the situation is different: in effect, the inability of destructive tests to take sure and certain account of weld quality has led to considerable development of non-destructive tests, so that it is to welding that non-destructive tests owe much of their progress. On the other hand, in the case of destructive tests, an arsenal of existing tests have been adapted to welding.

Destructive tests

Tensile tests

The arrangement most often proposed for tensile testing of flat assemblies with a fusion butt weld is that consisting of taking a flat tensile test specimen (or more rarely a round test specimen), perpendicular to the weld, situated in the centre of the calibrated section, the reinforcement being removed from the parent metal.[1] When such a test specimen is subjected to tensile testing until it breaks, the break appears (Fig. 10.1) either in the weld zone (case a) or in the parent metal (case b). If the break occurs in the weld zone, we can see a fracture load, but the value of that load depends, for a given test specimen width L, on width l of the weld zone. By a consolidation effect due to neighbouring parent metal, deformation diminishes and the load increases when the width of the weld zone diminishes.[2] Also, it is impossible to measure elongation because the extent of the deformed zone is insufficient for a significant result to be noted. If the fracture occurs in the parent metal (case b), we can obviously note a fracture load value in that metal,[3] whilst the elongation measurement is dubious because of the stiffening created by the weld bead in the calibrated section of the test specimen. As for the weld zone, we learn nothing about it. Finally let us note that we can observe the beginnings of reduction in area in the parent metal, on the unbroken side, symmetrical to the broken zone, compared with the weld metal.

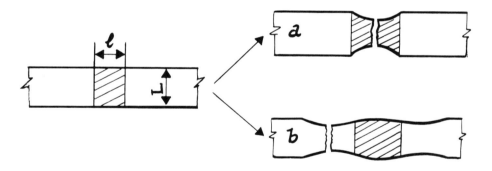

10.1 Transverse tensile test on flat butt welded assembly: a) Fracture with localised reduction in area in the weld zone; b) Fracture in the parent metal with deformation heterogeneity in the weld zone.

These comments very much restrict the scope of the transverse tensile test and explain the prudence of the relevant text in Standard NF A89-200, in which it states simply that the fracture load should be reported and position of the fracture noted.

Faced with these disadvantages, we could, still for flat assemblies, also consider taking and testing tensile test specimens in which the weld lies along the longitudinal axis of the test specimen. Thus, when carrying out the test, all the points of a section of the calibrated part are subject to the same strain. But this homogeneity of plastification is maintained only until initiation of a fracture at a point in the section, for example in the weld (Fig. 10.2) the fracture which gives rise to tearing of neighbouring zones, with different deformation. This is why the longitudinal test specimen is not normally used for tensile tests[4] but only, as we shall see later, for the bend test (French standard A89-204).

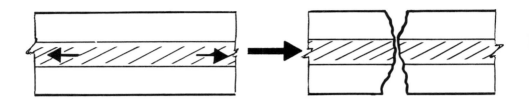

10.2 Mode of fracture of a longitudinal tensile test specimen on flat butt assembly.

Finally, the sole use which can be made of the tensile test, supplying the data which can normally be expected (strength, yield strength, elongation and reduction in area) is on a test specimen taken longitudinally from the weld metal of flat assemblies (Fig. 10.3) (French standard A89-302).

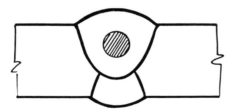

10.3 Taking a longitudinal tensile test specimen from the weld metal.

As we have already emphasised (see Chapter 4), this application of the tensile test should not be confused with that used for specifying filler products for arc welding and which relates to metal deposits carried out under conventional conditions and thus not representative of an actual assembly, as is the case here.

Bend tests

The application of standard bend tests to welded assemblies also presents difficulties associated with greater or lesser heterogeneity of assemblies to be tested. Nevertheless, because of its simplicity and the possibilities it offers in the evaluation of weld quality, the bend test is widely used, especially on transverse test specimens, less frequently on longitudinal test specimens. In both cases, reinforcement and the penetration bead are removed because their presence makes the test difficult to carry out.

The transverse bend test (French standard A89-203) is performed in such a way (Fig. 10.4) that either the face (a), reverse side (b)[5] or surface of a section (c) of the assembly under test is put under tension. The bend angle imposed is 180°; once this angle has been achieved, an examination is made of the appearance of the stretched section where any defects will arise. The dimensions of the mechanism used (chuck diameter and distance between supports) depends on the thickness of the test specimen, which will be that of the assembled thicknesses or of the sample taken. In fact, the test only makes sense if the elongation at the stretched section is more or less uniform in the fusion zone and neighbouring parent metal. Where this is not so, one of the two is being excessively deformed compared with the other (Fig. 10.5).

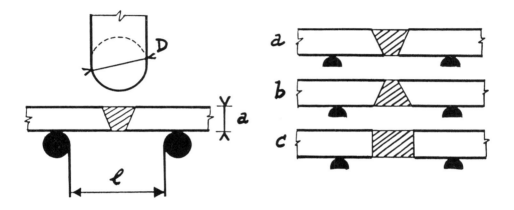

10.4 Transverse bend test: a) On the reverse side; b) On the face; c) On the side (D = 4e l = 6.2e).

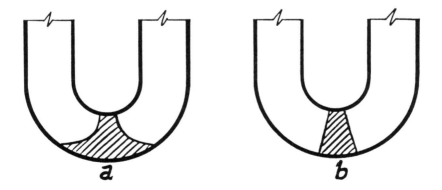

10.5 Results of transverse bend test: a) Weld zone more deformed than parent metal; b) Parent metal more deformed than weld zone.

It is to remedy this situation that Standard NF A89-204, in the case of such dissimilarity of elongation between weld and parent metal, proposes that the bend test be performed on a longitudinal test specimen, i.e. where the fusion zone occupies the centre line. Nothing else is changed in the test conditions, except of course that the side test is not proposed.

The difficulties resulting from the effect of local heterogeneity of welded assemblies during tensile or bend tests have led to various arrangements which we can usefully mention here without spending too much time, for they also have disadvantages. They have consisted of localising the deformation and fracture at the spot required, most often the weld zone, by reducing the test specimen section there (Fig. 10.6) by means of various contrivances with a zone where thickness and/or width are reduced. Such contrivances may prove useful for comparative purposes (for filler products and welding procedures), but they do not supply quantitative data likely to appear in standards or regulations or even less be taken into account in calculations.

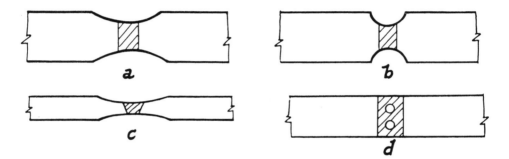

10.6 Examples of reduced section test specimens in weld zone: a) Reduced width, progressive curve; b) Reduced width, constant curve; c) Reduced width (called reduced volume test specimen); d) Section reduced by two round holes.

Another method, from Chevenard* and recently taken up by other authors, permits local determinations by means of micro test specimens (diameter of the order of 1.5 to 2 mm) which, tested on a machine specially developed by the author have made it possible to establish the first plotting of properties across welded joints, under traction (measurement of R, Re and A%) and on shearing. In particular, a series of micro-shear tests can be carried out on a single test specimen (Fig. 10.7) and the results entered on the macrographic image in correlation with the hardness plotting.

Notch impact toughness test

The test which remains the most used for evaluation of susceptibility to brittle fracture at a welded joint is the notch impact toughness test,

---

* P CHEVENARD: Study of the mechanical properties of welds by micromachine – SIS Bulletin, May – June 1935, no. 54, pp. 1760–72

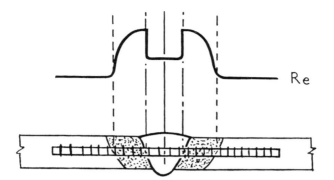

10.7 Diagram of shear strength relationship on micro-test specimen (according to Chévenard).

which has the advantage of being simple and which, applied to a notched test specimen[6] of small dimensions, offers the means of localised determination and thus of making comparisons on the basis of the position of the notch in the weld being tested. Furthermore, the test can be easily carried out at temperatures below ambient which makes it possible to call on the notion transition temperature. Currently, in France as abroad, steels are specified on the basis of notch impact toughness value (Charpy V) at graded temperatures. This practice has spread to welding whether it is a matter (as we have already seen in Chapter 4) of filler product qualification or tests on assemblies.

Assembly tests are governed by French standard A89-202 (still for flat and tubular assemblies) which specifies the arrangements (re thickness and distance) to be made for positioning the notch at the desired spot on the assembly. With regard to these arrangements, we can make the following observations:

a) In principle, the bottom of the notch should be perpendicular to the assembly surfaces (i.e. the rolling surfaces in the case of a rolled product). For this position (Fig. 10.8a), unless the fusion boundary and isotherm are themselves perpendicular to the surfaces, the fracture affects, from the bottom of the notch, an area in which the structure is not homogeneous. The designation of this position is therefore purely conventional; this is also often the case in the weld metal (Fig. 10.8b).

b) Standard NF A89-202 specifies the position[7] where the person carrying out the test is obliged to situate the bottom of the notch parallel to the surface (Fig. 10.9). In actual fact, this arrangement does not avoid the disadvantage mentioned above. Furthermore, it prohibits all comparison with the impact toughness of the parent metal, which is always determined on a notch perpendicular to the skin.

c) To permit the impact toughness test to be extended to assemblies with thicknesses incompatible with the normal 10 × 10 mm test specimen size, the standard allows for use of test specimens of reduced section (5 mm and 7.5 mm widths), the fracture energy being ascribed to the broken section, as with a normal test specimen. Here again, there is no possible comparison with the impact toughness

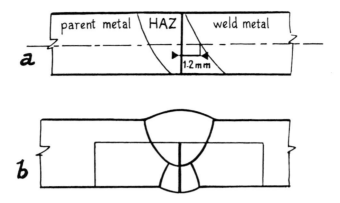

10.8 Conventional notch positions for measurement of impact toughness (French standard A89-202): a) In the heat affected zone and fusion boundary; b) In the weld zone.

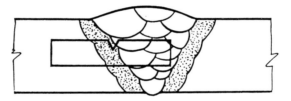

10.9 Position of an impact toughness test specimen with notch parallel to the surface for testing in the affected zone.

of the parent metal, unless a comparative determination is carried out on a geometrically identical test specimen.[8]

Role of fracture mechanics

Valuable as it is in evaluating and comparing the different zones of a welded assembly from the point of view of brittle fracture risk, the impact toughness test does not supply a quantitative response to the problem of the performance of that assembly when it harbours a defect from which a break, in particular a brittle fracture, may initiate. Indeed, the result of an impact test on a notched bar is expressed by an energy value (KV in joules or KCV in joules/cm^2 which cumulates the test specimen deformation energy prior to fracture and the actual fracture energy.[9] So we get no information on the conditions in which a break initiates, in the item under test, at right angles to a defect that that item may contain.

This is indeed the problem confronted with welded assemblies when, for purposes of establishing inspection procedure, a maximum acceptable dimension of defects in relation to anticipated service stress has to be set. The approach called 'fracture mechanics' supplies a solution which makes it possible to link the dimension of a flat defect such as a crack and the fracture toughness of the material, on the one hand, and the loading of the stressed component on the other hand (Fig. 10.10). To that end, amongst other criteria a stress intensity factor which is connected by a formula of the type: $K = \alpha\sigma \sqrt{\pi a}$ to stress σ and the dimension[10] α of the defect (comprising a fine crack) has been defined, α being a coefficient relative to the geometry of the test specimen (Fig. 10.10). From the stress measured at the break, we can deduce a critical factor symbolised by K_{lc}, which expresses the toughness of the material tested. But this formulation, called linear mechanics is set up on theoretical bases which presuppose limited deformation at the base of the crack, which implies either a high yield strength, or an important thickness. If such is not the case, i.e. if the deformation at the bottom of the notch exceeds the validity limits of linear mechanics, another approach consists of taking the deformation itself as a criterion of measurements of toughness, in so far as it has been possible to connect it to stress by means of semi-rule of thumb formulae in which theoretical uncertainty has been reduced by experimental input. Such is the basis of elasto-plastic mechanics which use the measurement of crack opening displacement (COD) symbolised by δ_c.[11]

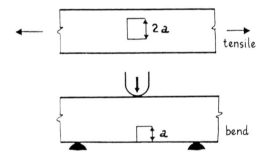

10.10 Taking account of dimension (a) of a flat defect during a fracture mechanics test under tensile and bend stress.

In France, the measurement of K_{lc} on homogeneous steel test specimens is covered by standard NF A03-180 but there is no standard on the measurement of COD. On the other hand, this measurement is standardised in many other countries, essentially on the basis of the British standard BS 5447.

As with the standard mechanical tests, there are a few difficulties and reservations in the application of fracture mechanics to welds. Indeed, to explore all of a, from the point of view of toughness, it has to be possible to locate the fine crack representing the defect where the break will initiate (usually a fatigue crack) in the various zones constituting it and this is not without difficulties, especially as the fracture may deviate into neighbouring regions. Also, the calculation of critical defect dimensions from δ_c takes account of the effect of residual stresses, implying the use of test specimens sufficiently large to remain representative of the actual assembly. Despite these difficulties, the literature mentions numerous practical applications and, lacking a French standard on the subject, we can usefully refer to the relevant British standard which refers to the measurement of the COD on welds or to the IIW recommendation on the same subject*.

Finally, it must be emphasised that tests derived from fracture mechanics are of interest only if carried out on an actual assembly, or on a check sample made by the same welding procedure. For qualification of filler products on a sample of deposited metal under conventional conditions, a test such as the measurement of the COD supplies no more useful information than the notch impact toughness test, but is much more complicated to carry out. Also, correlations have been established between impact toughness tests and fracture mechanics tests (at least for linear mechanics), which for this type of determination could be used on weld metal, without recourse to other tests.

Resistance spot welds

In actual fact, none of the techniques offered by the range of standard mechanical tests can be applied directly to resistance spot welds. Furthermore, tests which have been devised and developed provide only an imperfect indication on service performance, because in the majority of cases, a spot weld is not designed to be stressed in isolation, but rather to contribute to the constitution of an assembly where several points work simultaneously.

Nevertheless, such tests are necessary for they permit comparison of results given by the various settings that a welding machine can provide, to verify the suitability for spot welding of products tested together with the maintenance of spot quality during series construction. To that end, the French Standards include the tensile shear test (French standard A89-306) and the torsion shear test (French standard A89-205).

The tensile shear test consists of subjecting a flat test specimen to traction until it breaks (Fig. 10.11a); this specimen includes the spot tested, compensating where necessary the misalignment in the jaws due to the thickness of the two assembled half-specimens (Fig. 10.11b) from 3 mm for steel. Despite this precaution, there is no clean shear because the test specimen deforms at the spot during the course of the test and the fracture occurs under the combined effect of a traction component associated with the shear stress. Apart from the cases of fractures involving the parent metal to a greater or lesser degree, the result is two aspects of fractures, one called shear fracture (involving the diametral plane of the weld zone

---

* IIW Doc. X 614-71: Recommended techniques for fracture toughness testing. Specialised working group progress report – 1971

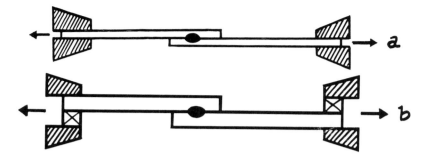

10.11 Tensile shear test on spot welded test specimen: a) With no compensation of thickness (e < 3 mm); b) With thickness compensation (e ⩾ 3 mm).

(Fig. 10.12a)) and the other slug or peel fracture, affecting the periphery of the weld zone which remains intact (Fig. 10.12b). The peel test consists of obtaining, without special material or measurement, one of these two aspects when the break is obtained by tearing. A good quality spot weld will break in accordance with mode b. This test has the advantage of being simple and lending itself readily to statistical monitoring of spot quality although it supplies little information of a metallurgical nature.

More complicated since it requires the use of a special machine, the torsion shear test supplies quantitative information on the intrinsic qualities of a spot weld (via the torque and angle of fracture), qualities which can be interpreted by examination of the fracture which, normally, occurs in the equatorial plane of the spot weld. In an as-welded spot weld, we can see more or less distinctly (Fig. 10.13), depending on the welding sequence, three zones which correspond to the process of spot formation as revealed by the macrograph and which we have already mentioned in Chapter 5, namely:

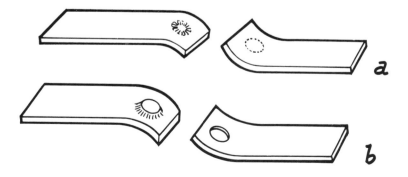

10.12 Appearance of tensile shear fractures of resistance spot weld: a) Shear fracture; b) Peel fracture.

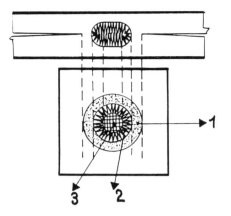

10.13 Macrographic interpretation of the appearance of a torsion shear test on a resistance spot weld: 1) Peripheral tear zone; 2) External ring of the weld zone (basaltic crystallisation); 3) Central area of the weld zone.

– peripheral zone 1 at the outer limit of the fusion zone, where the rise in temperature and pressure have permitted a more or less complete solid phase pressure weld. Operators attribute much importance to the existence of this zone (called peripheral tearing) in regard to spot weld quality.

– in the weld zone, a ring 2 corresponding to dendritic solidification progressing from the exterior towards the interior of the spot weld (like the subsequent transformation).

– and thirdly, in the central part, an area 3 which corresponds with the convergence towards the diametral plane of the solidification grains arising from the periphery and, eventually with the presence of an equi-axial structure. This is the zone in which, in addition to the effects of this process of solidification, we can also see the effects of any segregation, sometimes with porosities.

Naturally, these three components of the aspect of fracture appear on the curves recorded; however their interpretation must take account of the normal, continuous variation of torque when the torsion angle varies with progression of the fracture. Also, in the case of steel, the appearance of the fracture is of course modified by heat treatment between electrode tips, as is the structure itself.

Non-destructive tests

Figure 10.14 contains the indications which can be given concerning the principal non-destructive testing procedures[12] (penetrant fluid, magnetic particle, radiographic, ultra-sound) used for examining welds, either from the point of view of the possibilities they offer for detection and, if possible,

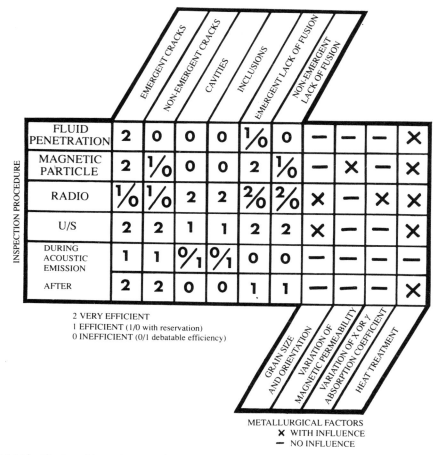

2 VERY EFFICIENT
1 EFFICIENT (1/0 with reservation)
0 INEFFICIENT (0/1 debatable efficiency)

METALLURGICAL FACTORS
✗ WITH INFLUENCE
— NO INFLUENCE

10.14 Classification of inspection procedures.

identification of defects, or from the point of view of any influence of certain metallurgical factors. For the terminology of defects, the first four headings proposed by French standard A89-230 have been used, without going into the detail of sub-divisions;[13] these are cracks, cavities, solid inclusions, lack of fusion defects, of which the first two concern defects resulting principally from a metallurgical process. Included as metallurgical factors were the structural elements of grain size and orientation and constitutional modifications which result in a variation of the ferromagnetism or X or γ ray absorption coefficient; in addition, account was taken of the fact that the component examined may or may not have undergone a post-welding heat treatment.

As regards detection and identification of defects, we can refer to the comments below the semi-quantitative classification proposed in Fig. 10.14.

We shall not dwell on the indications relative to inspection by fluid penetration or magnetic particle procedures. We shall simply note that the reservation relative to detection of lack of fusion by fluid penetration concerns the possibility that the defect may be filled by slag or oxide. For magnetic particle inspection, the reservation relative to the detection of non-emergent cracks or lack of fusion results from the influence of the relative height of the defect in relation to the thickness of the component.

With radiographic inspection, a procedure much indicated for the detection of cavities and inclusions, the reservations concern the fact that a relatively thin, flat defect such as a crack or lack of fusion, may be traversed and not covered by the radiation, which means it escapes detection. This situation is more frequent for cracks than lack of fusion (with oxide inclusions), hence the difference in assessment.

According to the figure, ultrasonic inspection seems to be the most indicated for detecting flat weld defects, provided that the operator is careful to ensure adequate variation of the beam angle in relation to the defects.

The indications relative to acoustic emission are given somewhat hesitantly because we do not have enough experience in this subject, which explains the lack of recommended practices. When carrying out acoustic monitoring of a component in process of being welded or during subsequent cooling, every effort must be made to filter out the spurious noises due to the welding itself (from the arc, solidification shrinkage and in the solid state, slags, etc) so that, for the moment, revealing inclusions and cavities during their formation, which seems possible, requires precautions not compatible with normal industrial practice. Crack detection seems more feasible, particularly for cold cracks which, due to the mechanism of their formation (delayed cracking), appear after the noises associated with the welding itself have disappeared. There is still the question of being able to localise cracks thus detected, which requires quite a sophisticated technique.

To carry out acoustic inspection after welding, i.e. on a finished piece, it is necessary to apply stress[14] for example by applying pressure in the case of a tank) which causes deformation where there are defects, mainly flat defects, hence a detectable and locatable acoustic emission, particularly for cracks but also for some lack of fusion. But, as things stand at present, acoustic inspection is not widely used.[15]

As for the metallurgical factors which play a role in non-destructive testing, mentioned in Fig. 10.14, we can explain their influence as follows:

– grain size and orientation play a role in X-ray and ultrasonic inspection: in radiographic inspection of aluminium or austenitic stainless steel fusion welds (for example), we can often see variations of opacity in the radiograph of the weld zone and sometimes the fusion zone, especially with automatic welds; these variations could be taken for defects. They are not defects, simply patches of X-ray diffraction on the crystallographic planes of solidification grains which, as we saw in Chapter 5 have a privileged growth direction, which is particularly marked in the case of fast welds. These diffraction patches, which resemble those of crystallographic powder diagrams, move apart when the distance between component and film is increased, which makes it possible to distinguish the true defects as the image of these does not move.

In ultrasonic inspection, it is essentially the size of the grains rather than their orientation which creates difficulties in welding single phase alloys such as austenitic steel. It is a resonance effect which spoils the spread of the ultrasonic beam and disturbs the indications, in terms of its frequency in relation to the size of the grains. Particular precautions must therefore be taken, notably in varying the frequency.

– an obvious cause of variation in magnetic permeability, a real nuisance in magnetic particle inspection in steel, lies in the use of an austenitic and therefore non-magnetic filler. The welded joints achieved cannot be subjected to this type of inspection because the leakage lines jump from one side to the other of the weld zone, giving a false image by accumulation of magnetic powder at the fusion boundary. To a lesser extent, for hardening steels, martensitic transformation of the affected zone gives rise to variations in magnetic permeability which also constitute sources of erroneous interpretation.

– when the welding effect results in the appearance of constituents creating a variation in absorption coefficient vis-a-vis ionising radiation, the corresponding X-rays mark their presence by areas of differing exposure. Such is the case, already mentioned, for resistance spot welds on aluminium alloys containing copper: separation of the $Cu\,Al_2$ constituent (which is more absorbent) at the periphery of the fusion zone, defines the limits of this and the corresponding image is the sign of a good quality spot.

– finally, for all the inspection procedures,[16] the figure summarises the possible effect of a post welding heat treatment, through the intermediary structure of stress relief. The structure effect is particularly sensitive in magnetic particle inspection, because tempering (or annealing) suppresses the heterogeneity mentioned above. The effect of stress relieving generally results in cracks opening (whether they emerge or not), which is an advantage for all traditional inspection procedures. In acoustic emission after welding, the advantage arises from the fact that disappearance of residual stresses makes diagnosis safer as the stress applied at the moment of testing is the only stress involved. Such are the reasons for which, when the manufacture of a welded assembly includes a final heat treatment, it is recommended that non-destructive testing be carried out after that heat treatment,[17] or at least that a verification of the results of any pre-treatment inspection be carried out at that time. Naturally, if there are reasons to fear the phenomenon of reheat cracking, the post-treatment inspection is absolutely necessary.

Chapter 10 **Comments**

1 If the reinforcement is maintained, the fracture in the weld zone is unlikely, and the result shows simply that the weld 'holds' better than the parent metal or not but does not establish that it is healthy.

2 This effect is sensitive for electron beam welds and still more for braze welds which may break elsewhere than in the weld zone, even if the weld zone is intrinsically weaker.

3 By localising the fracture it causes, we can also reveal any softening effect of the parent metal due for example to recrystallisation or over-tempering. But, depending on the width of the zone in question, this figure gives rise to the same reservations as those expressed above in the case of fracture in the weld zone. The more narrow the softened zone, the more the consolidation effect arising from the fracture load is involved. So, for example, an EB weld will fracture in the parent metal for a higher load than a TIG weld.

4 However, the test carried out without removal of the reinforcement takes account of the practical performance of welds working principally in the longitudinal direction (for example circular welds on liquid gas cylinders).

5 Or indeed each of the weld surfaces tested in the case of symmetrical preparation.

6 To distinguish between tests on U and V test specimens, the French standard concerning the test on a V notch (French standard A03-161) describes it as 'impact bend test on a doubly-supported V notched test specimen'. In everyday language, we talk about 'Charpy U test' and 'Charpy V test'.

7 Obligation which can only arise from an individual specification, but not for a generally accepted technical reason.

8 However, some unofficial documents provide correlations.

9 These two values may nonetheless be distinguished and measured separately by means of special techniques (instrumented dropweight tests).

10 The word 'size' used here designates the important defect dimension vis-a-vis the initiation of the fracture, namely the length in the case of a tensile test specimen or depth for a bend test specimen (Fig. 10.10).

11 Like impact toughness, value δ_c is dependent on temperature. So we can define a ductile-brittle transition temperature.

12 To which has been added acoustic emission, which is not a widespread inspection procedure but which can usefully be mentioned because this procedure can be applied during welding or after; during welding it permits detection of defects whilst they are forming and analysis of the process by which they appear.

13 In examining this detail in the standard mentioned, we can easily see that the validity of the general indications in this figure would need to be verified in each individual case.

14 This technique is not to be confused with that of leak detection which is carried out by listening to the noise caused by the leaks themselves.

15 Acoustic emission does however belong to the procedures taken into account by the COFREND Certification Committee.

16 Except, by definition, in the case of acoustic monitoring during performance of a weld.

17 An inspection totally carried out after treatment is not without disadvantages because, in the event of rewelding any defects noted, repeated heat treatment may then be necessary, if not a further inspection.

Index

This index has been compiled to make use of the book easier by complementing the contents list. Consequently, with a few exceptions, concepts already expressed in the contents have not been included. Similarly the reader will not find in this alphabetical list the names of all the welding processes mentioned in the text. Only processes such as electron beam, spot or electroslag welding are included because of the individual characteristics they present in relation to other processes.